Single-Pilot CRM

Phil Croucher

About the Author

Phil Croucher holds JAR, UK and Canadian professional licences for aeroplanes and helicopters and around 7600 hours on 37 types, with a considerable operational background, and training experience from the computer industry. He has at various times been a Chief Pilot, Ops Manager, Training Captain and Type Rating Examiner for several companies. He is currently Chief Pilot of the Western Power Distribution Helicopter Unit in the UK and is responsible for the safety column in *Vertical* magazine. He can be contacted at **www.electrocution.com**

Legal Bit

Copyrights, etc.

All Rights Reserved

Praise for JAR Professional Pilot Studies:

"I've made pretty good progress with your book and I'm really impressed. I've found it readable and really pertinent to the ATPL exams."

Balbir

Praise for The Helicopter Pilot's Handbook:

"a much needed summary of all the "basics" we live by. It sure is great for an old timer to see how all the things we had to find out the hard way in the 60's and 70's can now be found in a book. A great reference tool."

Robert Eschauzier

"A good insight into the working life of a pilot"

HF

"In short... great book! Thankfully your book illuminates many of the practical aspects of flying rotorcraft that are missing from the intro texts used during training. I'd equate much of the valuable practical information in your book to the to the same valuable information found in "Stick and Rudder."

Ian Campbell

".....an excellent book for pilots interested in a career as a helicopter pilot. It answers all the really hard questions like "how does a young pilot get the required experience without having to join the army for 10 years". Great book for anyone interested in fling-wings!"

Reilly Burke, Technical Adviser, Aero Training Products

"Having only completed 20 hours of my CPL(H) in Australia, a lot of the content was very new to me. Your writing style is very clear and flowing, and the content was easy to understand. It's made me more eager than ever to finish my training and get into it. It's also opened my eyes as to how much there is to learn. The section on landing a job was excellent, especially for this industry that seems so hard to break into."

Philip Shelper

"Picked up The Helicopter Pilot's Handbook on Friday and have already read it twice. How you crammed that much very informative info into 178 pages is totally beyond me. WELL DONE. What a wealth of information, even though I only have a CPL-F. OUTSTANDING. I'm starting it again for the third time because I've picked up so much more the second time, that I'll read certainly a dozen more times. I cant wait to apply a lot the ideas and comments that you have supplied.

My wife is totally blown away that I've read it cover to cover twice and going around for a third time. She said it must be an outstanding book as I need real mental stimulus to keep me going.

Will

".....provides many insights that wouldn't appear in the standard textbooks.

The next part of the book deals with the specialised tasks that a jobbing pilot may be called upon to do. It covers a very wide range of tasks from Avalanche Control through Ariel Photography and Filming; from Wildlife Capture to Winter Operations; from Pipeline Survey to Dropping Parachutists. This is not an exhaustive list of what he covers. If you need to know then it's probably here. The information given is good, practical and down to earth. It is exactly what you need to know and written from the pilot's point of view.

For anyone with some practical experience of helicopter operations it is worth a read. For someone who is going into civvie street and intends to fly then it is definitely worth a read. For anyone who intends to be a 'jobbing pilot' it could be invaluable as a source of reference."

Colin Morley
Army Air Corps

"Your book is very good and has been read by a few of the guys here with good 'raps'. particularly the Info on slinging etc. is stuff that is never covered in endorsement training. Certainly a worthwhile addition to any pilot library."

Gibbo

"I have only skimmed through the first version. Its already answered and confirmed a few things for me. Just the type of info I am after."

Andrew Harrison

Praise For The BIOS Companion

"The computer book of the month is The Bios Companion by Phil Croucher. Long-time readers of this column will recall I have recommended his book before. This tells you everything you ought to know about the BIOS in your system. Post codes, options, upgrades, you name it. Years ago, I called an earlier edition of this invaluable and I see no reason to change my view. Recommended."

Jerry Pournelle, *Byte Magazine*

"You will find more information about your motherboard assembled here than I have ever seen."

Frank Latchford *PCCT*

"Thank! I really appreciated this. I read it and was able to adjust my BIOS settings so that my machine runs about twice as fast. Pretty impressive. Thanks again."

Tony

"This book is worth far more than is charged for it. Very well written. Probably the most-used reference book in my shop. .....a great value as the feature explanations trigger your thinking and allow you to figure out many related BIOS features in some of the newer versions."

Amazon reader

"I received my package today containing the BIOS Companion book and 2 CD set.... I'm really impressed with what I did receive. I already had about HALF of the information, and to get THAT much, I had to get several books and web pages. GOOD JOB!!

I had more time to go thru the book and think that you should change the word "HALF" to "FOURTH".

I commend you on the great job you did. That's a hell of a lot of work for any major company to do, let alone an individual."

Craig Stubbs

Contents

CONTENTS

CONTENTS

INTRODUCTION

Although the title of this book implies otherwise, its real thrust is Decision Making (together with an appreciation of human factors), since this is the area that most affects pilots working by themselves - CRM's function, in the guise of better crew interaction, is actually to facilitate the decision making process, but the popular conception is the opposite, hence the relative positioning of the two ideas in the title. All the reader needs to be aware of is that the two terms are used synonymously in this book.

Another problem is that the sort of mistakes that cause accidents do not arise directly from situations where CRM is relevant, but from within individual pilots - if you want to be technical, they arise from *intrapersonal* rather than *interpersonal* causes. Modern life is stressful enough - we are all hostages to other peoples' expectations and attitudes, and it often seems that, within an hour of waking up, we have an attitude all of our own, by the time you've dropped the toast on the floor and everyone's had their bite out of you. It is very important not to let what happens outside into the cockpit - for example, one pilot ran out of fuel in a Longranger by trying to collect a fire crew when he really should have been off to the refuelling point (there were plenty of other machines around to do the job). It later turned out that he was always running his car low on fuel, and it was part of his personal make up. If you like, he was accident-prone before he started (this is a typical result of *false hypothesis*, where you get a wrong idea in your head, which especially occurs when low on fuel as a strange sort of denial sets in).

Also, it has (finally) been realised that traditional methods of flight instruction have been missing something - the assumption has always been that, just because you have a licence, you know what you are doing, or that good, technically qualified pilots made good decisions as a matter of course. Everybody on the shop floor, of course, has always known that this is not necessarily so, and a lot of experienced pilots make mistakes, so experience is not the answer, either. In fact, experience can be a harsh teacher, assuming you heed its lessons anyway, so ways have had to be found to use training instead, hence CRM, or the more appropriate Pilot Decision Making, as it's called in Canada.

The current thinking sees aeronautical decision making as a function that is open to analysis under standard psychological theory and practice (*Brecke*, 1982; *Stokes and Kite*, 1994). Research into the human factors related to aircraft accidents and incidents has highlighted decision making as a crucial element (*Jensen*, 1982; *O'Hare, Wiggins, Batt, and Morrison*, 1994).

The irony is that people who are aware that CRM is a Good Thing do not need CRM courses - the sort that should most benefit are like the Enstrom owner who mentioned to his shocked engineer that he didn't like the look of two bolts in the tail rotor assembly, so he turned them round and shortened one of them, since it was longer than the other. After patiently explaining during wall-to-wall counselling that the reason why one bolt was longer was for balance purposes, and that they were inserted a particular way round for a reason, the engineer suggested the owner-pilot took his custom elsewhere.

Or there's the one who was asked by an examiner what he would do if an engine failed at low level. The reply included the statement: "I would go into cloud to gain some height." On being taken into cloud and shown what icing looked like, the response was "Oh, we get that all the time!"

In the film *An Officer & A Gentleman*, a gunnery sergeant asks why a hustler would want to go through boot camp, to which the hustler replies "I want to fly jets, Sir!". The sergeant's response is: "My granny wants to fly jets - we're talking about character!" So how do you teach character, which is really something that can only be developed in concert with other people?

This book is a modest attempt, and, at times, may seem a little "New Age" in its outlook, especially Chapter 5 - all I can say to that is that many doctors now think that most of the body's ills are psychosomatic, meaning that they stem from mental courses (which explains miraculous recoveries), and even such hard-nosed people as the New York Police have recognised that people seem to attract unwanted events into their lives, hence their saying "There are no victims". What they mean is, random external events don't just happen, and that people attract them according to what they expect out of life, and are completely responsible for what happens to them.

ACCIDENTS

The human factor is not everything - the "safety record" of an airline can be nothing but a numbers game. Take a flight from Los Angeles to New York with two hundred passengers on board - the distance is 3000 miles, so they have flown 600,000 passenger-seat miles. With 150 on the flight back, you get 1,050,000, for being in the air for only 9 hours! If they have 20 aircraft doing that five days a week, and injure one passenger, they can say that that happened only once in 105,000,000 passenger-seat-miles, which is still only 900 hours.

Aircraft are getting more reliable so, in theory at least, accidents should happen less often. Unfortunately, this is not the case, so we need to look somewhere else for the causes. Believe it or not, accidents are very carefully planned - it's just that the results are very different from those expected! And if you thought safety was expensive, think about these consequences of an accident:

- Fatalities and/or injuries

- Customer relations & company reputation suffer

- You need alternative equipment.......

- ....while still making payments on the one you just crashed

- Any schedule gets screwed up

- The insurance is increased

As with most other things, aviation is more of a mental process than a physical one. For example, it takes much longer to become a captain than it does to become a pilot, and CRM/PDM training aims to shorten this gap by substituting training for experience. Almost the first thing you have to take on board is that not everyone does things the same way as you do, as a result of which, compromises have to be made in order to get the job done. Another is that, when single pilot, feedback is missing, which is useful, when multi-crew, for making decisions. The only real replacement for this is reviewing your flights and discussing them with colleagues, which is more difficult for helicopter pilots, because of the lack of flying clubs and other meeting places (but licensed premises are good).

Particularly risky situations include trying to beat the weather, not having enough fuel and flying when ill, not to mention fatigue. Another factor is the unclear role of the pilot - just about everybody in authority tells you that you are in charge, and to fly safe, etc., but the reality is that you are either a glorified bus driver, chauffeur or taxi driver, subject to the whims of employers, and are often expected to falsify your duty hours and tech logs, fly in bad weather, or overweight, or both, as a matter of routine. Stress can also arise from working with inadequate equipment, or people, such as when Those On High in your company are not good leaders or communicators, which is hardly surprising, as most of them have never had the proper training. Most of aviation management comes straight from the ranks, and they might be good pilots, but managers?

An accident is actually the end product of a chain of events, so if you can recognise the sequence it should be possible to nip any problems in the bud (see elsewhere in the book about Professor Reason's Swiss cheese analogy). Human factors training is supposed to help you recognise the events that lead to errors (and accidents) and to avoid those situations as much as possible, but it is only really effective if the whole system is designed to be error-resistant, which is where your company's Quality System comes in, to ensure safe operational practices and airworthy aeroplanes. To achieve this, the system is documented and inspected periodically by an auditor who uses a series of checklists to detect non-compliances, which are reported back to a Quality Assurance Manager for rectification. Although a certain amount of documentation already exists, in the form of Operations Manuals or MMEs, for example, the Quality System extends things by introducing compliance monitoring and a feedback system to provide confidence that the whole system is working. However, aside from checking their paperwork more, about the only thing the average pilot would notice is that, if anything untoward happens, a form is filled in.

HISTORY

A common saying is that "the well oiled nut behind the wheel is the most dangerous part of any car", which is not necessarily true for aviation, perhaps (aside from the "sloppy link" between the controls!), but, in looking for causes other than the hardware when it comes to accidents, it's hard not to focus on the pilot (or other people - e.g. the human factor) as the weak link in the chain - around 75% of accidents can be attributed to this, though it's also true to say that the situations some aircraft are put into make them liable to misfortune as a matter of course, particularly with helicopters - if you continually land on slippery logs, something untoward is bound to happen sometime! The reason why 747s don't have accidents in forests is because they don't fly near them! Similarly, the reason why so many experienced pilots have heliskiing accidents is because only experienced pilots are employed on it!

The trend towards human factors in relation to accidents was recognised in '79 and '80, where over 500 incidents relating to shipping were analysed, and 55% were found to be related to human factors. Did you think that was *1979* & 80? It was actually in *1879* and 80! Since then, through the *1*980s and 90s, aviation accidents in the USA were analysed in depth, and it was found that *crew interaction* was a major factor in them since, nearly 75% of the time, it was the first time they had flown together, and nearly half were on the first leg, in situations where there was pressure from the schedule (over 50%) and late on in the duty cycle, so fatigue was significant (doesn't everything happen late on Friday afternoon?) The Captain was also flying 80% of the time. The problem is, it's not much different now - 70% of accidents in the USA in 2000 were pilot-related, based on mistakes that could easily be avoided with a little forethought (it was more or less the same figure way back in 1940!). Now, the figure worldwide is around 80%. If air traffic continues to grow at the present rate, we will be losing 1 airliner per week by 2010, and even more GA aircraft - the Australian authorities are looking at 1 helicopter per week.

Since the problem of crew co-operation had to be addressed, management principles from other industries, such as Quality Assurance and Risk Management, were distilled into what is mostly called *Crew Resource Management*, triggered, in Canada, at least, by three accidents, one of which was at Dryden, which was also instrumental in new Canadian icing laws being passed (the Captain tried to rotate an ice-covered aircraft twice, but only made it into the trees - the machine had to be refuelled with an engine running as the APU was unserviceable and there was no external power unit, which meant that deicing was prohibited; all this on a behind-schedule day. Although he bears the final responsibility, the Captain sure didn't get much help from elsewhere). In fact, most weather-based accidents in small aircraft involve inadvertent entry into IMC by people with only the basic instrument stuff required for the commercial licence. Next in line is icing. With regard to jet transports and executive jets, it's CFIT, and the figures are 50% and 72%, respectively.

What we now know as CRM was actually developed from the insights gained after installing Flight Data Recorders and Cockpit Voice Recorders, when crews were not considered to be assertive enough, and Captains not receptive enough (CRM in those days could be summed up with the phrase "I'm the captain - you're not!")

Prompted by a NASA workshop in 1979, United Airlines started to include the training, and not just for pilots. The goal was *synergism*, meaning that the total performance of a crew should be greater than the sum of its parts, or each crew member (like Simon & Garfunkel, or Lennon & McCartney, who are talented enough by themselves, but so much better as part of a group). To achieve this, members of any team must feel that they and their opinions are valued, and understand their roles. Since, in most companies, the teams change from day to day (or flight to flight), the whole organisation must foster teamwork, *from the top down*, and attempt to reduce the effects of jagged edges between people (in other words, the relatively simple concept of learning to live with others, which involves sharing power on the flight deck, at the very least). The behaviour of people in a company is very much a reflection of the management, so there is an obligation to foster a positive working environment, which, essentially, means not being surly or miserable - the company culture should allow anyone to speak up if they feel they have to. Like it or not, you are part of a team, even if you are the only one in the cockpit, and *you* have to fit into an established system (especially when IFR).

The concept evolved from the original *Cockpit* Resource Management, through *Crew* Resource Management, where Decision Making became more important, into a third generation, which began to involve cabin crews, etc, and introduce aviation-specific training, as a lot of what went previously was regarded as just so much "psycho-babble", but it is very difficult to escape psychology in just about every walk of life these days, and aviation is no exception - all airlines use selection tests. In fact, 90% of aviation casualties in World War I were directly attributable to human factors, and in World War II they started testing to weed out people who had questionable decision-making skills.

CRM then became integrated into all flight training, and an element is now met on nearly all check rides, with a complete syllabus cycle taking place over three years. In the US, the fourth generation can take the form of an *Advanced Qualification Program* (AQP) tailored specifically to individual company needs. Now we are in the fifth generation, which attempts to become universal and cover national culture, and concentrate more on *error management*, which accepts that sh*t happens, and adopts a non-punitive approach to it (which does *not* mean you should break the rules on purpose!) Evidence of this can be seen with anonymous reporting procedures, such as CHIRP in UK and the Aviation Safety Action Program (ASAP) in the US.

A further development could be to change the name yet again to *Company* Resource Management, where other departments aside from the flight crew get involved in the same training. The benefit of this for Air Aurigny (in the Channel Islands) has been improved communication between departments and a sharpening up of the whole operation once people saw what everybody else had to cope with - turnaround times became shorter, which made a direct contribution to the bottom line.

However, the general principles of CRM have been around for some time - Field-Marshal Montgomery wrote that the best way to gain a cohesive fighting force was efficient *management* of its components, and he certainly succeeded in getting the Army, Navy and Air Force to work together. However, as far as definitions go, I guess you could call it *Cockpit* Resource Management when you're single pilot, and *Crew* Resource Management when you're not.

Previously, you might have been introduced to the concept of *Airmanship*, which involved many things, such as exercising judgement, looking out for fellow pilots, doing a professional job, not flying directly over aircraft, etc. - something that could be called being the "gentleman aviator", or using your common sense. These days, those terms are not adequate, and there are new concepts to consider, such as *delegation, communication, monitoring* and *prioritisation*, although they will have varying degrees of importance in a single-pilot environment.

In fact, the term "pilot error" is probably only accurate about a third of the time; all it really does is indicate where a breakdown occurred, and the term normally includes direct mistakes, failure to follow SOPs, inappropriate judgement or failure to intervene. There may also have been just too much input for one person to cope with, which is not necessarily error, because no identifiable mistakes were made. Perhaps we need a new phrase that occupies the same position that "not proven" does in the Scottish Legal System (that is, somewhere between Guilty and Not Guilty).

Airmanship, by the way, is now referred to as *non-technical skills*.

WHY THIS TRAINING?

The aim is to increase flight safety by showing you how to make the best use of resources available to you, which include your own body (physical and psychological factors), information, equipment and other people (including passengers and ATC), whether in flight or on the ground, even using the humble map. Using only a GPS for navigation, and ignoring the other navaids or the map, for example, is a bad use of your resources.

WHAT'S IN IT FOR ME?

You should be able to make better decisions after being introduced to Decision Making. It has been noticed that pilots who receive such training outperform others in flight tests and make 10-15% fewer bad decisions, and the results improve with its comprehensiveness. However, remember that training cannot cater for everything. Instead, as with licences everywhere, you are given enough to be able to sort things out for yourself, or given enough rope to hang yourself!

So, we know all about the hardware, now it's time to take a look at ourselves. An *accident-prone person*, officially, is *somebody to whom things happen at a higher rate than could be statistically expected by chance alone*. Taking calculated risks is completely different from taking chances. Know your capabilities, and your limits. Things that can affect performance include:

- *Physical fitness* and suitability (visual acuity, etc)

- *Knowledge* - for example, know the flight manual, and its limitations

- *Preparation* - do as much as you can before the flight (this is part of workload management) - check the weather and performance. Is that runway *really* large enough? Can you stop on it if the other engine goes as well? Has all the servicing been done? Is the paperwork correct? Visualise the route from the map - and fold it as best you can for the route

- *Attitude, Personality* (Awareness, Discipline, Professionalism)

- *Experience & Recency*

- *Skills & Ability*

- *Arousal level, Alertness*

- *Cultural influences*

- *Environmnt (noisy, dark, cold hangars)*

- *Planning*

All the above help reduce the workload. After the saturation point, though, work must be *prioritised* to cope, and we are, in fact, more productive when we do things one at a time, rather than try to multitask, although it's true to say that we can do some things at once, like talk and drive, but not others.

CRM/PDM courses are supposed to be discussion-based, which means that you are expected to participate, with the intention that your experiences will be spread around to other crews. This is because it's quite possible never to see people from one year to the next in a lot of organisations, particularly large ones, and helicopter pilots in particular have no flying clubs, or at least opportunities to "hangar fly", so experience is not being passed on. In fact, if you operate in the bush, you might see some of your colleagues during training at the start of the season, and not see them until the end, if at all. Since your "crew" is mostly non-existent, the material is adapted slightly in this book to concentrate on decision-making and some stuff that is relevant to single-pilot operations. Having said that, of course, you still have to talk to management and engineers, and to people even more important - the customers (a CRM environment is particularly useful with heliskiing guides or jughounds!)

One accident which illustrates the need for CRM training was a Lockheed 1011 that flew into the Florida Everglades. A problem involving the nosewheel (actually one of the green lights) occupied the attention of all three members of the crew so much that they lost the big picture, and the aircraft ended up in the swamp. It was concluded that the commander should have ensured that someone was monitoring the situation, and delegated tasks accordingly. But was a "mistake" actually made? Nobody pushed the wrong switches or carried out the "wrong" actions. A contribution to the Kegworth accident in UK, where a 737 ended up on the motorway, was the inability of the cabin crews to feel they were able to talk to the flight deck if they saw a problem, which puts the problem fairly and squarely at the door of the Company, or at least the management. Also, the accident report on the Air Florida flight that hit a bridge and ended up in the Potomac is instructive - the FO was clearly sure that something was wrong (icing) but didn't like to say so.

In short, CRM (whatever the C stands for) is the effective utilisation of all available resources (e.g. crew members, aeroplane systems and supporting facilities) to achieve safe and efficient operation, by enhancing your communication and management skills. In other words, the emphasis is placed on the non-technical aspects of flight crew performance (the so-called *softer skills*) which are not part of the flying course and which are also needed to do your job properly - those associated with teamwork, and helping a team work together. As we said before, you could loosely call it airmanship, but the term *Captaincy* is more appropriate, as flying is way more complex than when the original term was coined.

CAPTAINCY

The elusive quality of Captaincy is probably best illustrated with an example, using the subject of the Critical Point. If you can think back to your pilot's exams, you will recall that it is a position where it takes as much time to go to your destination as it does to return to where you came from, so you can deal with emergencies in the quickest time. In a typical pilot's exam, you will be given the departure and destination points, the wind velocity and other relevant information and be asked to calculate the CP along with the PNR (*Point of No Return*), which is OK as far as it goes, but tells you nothing about your qualities as a Captain, however much it may demonstrate your technical abilities as a pilot.

Now take the same question, but introduce the scenario of a flight across the Atlantic, during which you are tapped on the shoulder by a hostess who tells you that a passenger has got appendicitis. First of all, you have to know that you need the CP, which is given to you already in the previous question. Then you find out that you are only 5 minutes away - technically, you should turn back, but is that really such a good decision? (Actually, it might not be, since it will take a few minutes to turn around anyway). Commercially, it would be disastrous, and here you find the difference between being a pilot and a Captain, or the men and the boys, and why CRM training is becoming so important. Put simply, you are not really a Captain until you can learn to say NO.

A Captain is supposed to exhibit qualities of loyalty to those above and below, courage, initiative and integrity, which are all part of the right personality - people have to *trust* you. This, unfortunately, means being patient and cheerful when all you want to do is rip somebody's face off, and even changing your own personality to provide harmony, since it's the objective to get passengers to their destination safely. As single crew, of course, there is only you in your cockpit, but you still have to talk to others, and we all work in the Air Transport Industry - it just happens that your company is paying your wages at the moment. In this context, the word "crew" includes anybody else who can help you deliver the end product, which is:

.. Safe Arrival!

Very few people travel just for the sake of it, unless they have a Pitts. These days, companies are there to make money, and not because the owner is an enthusiast and into aviation for its own sake, as it might have been in the old days. Everything else is subordinate to this, including pride and increasing your qualifications and experience. Remember that the general public are ultimately paying your wages.

The best way out of trouble is not to get into it, which is easier said than done with an intimidating passenger or management. You, the pilot, are the decision-maker - in fact, under the Chicago Convention (Annex 2, Chapter 2), your word is law until overturned within 3 months by a person with a lawful interest. However, the other side of the coin is that you are liable for what goes on - in fact, in aviation, the buck stops right at the bottom. The complication is that, although many companies say "safety first", it all goes out of the window when a customer is waiting or problems arise. In reality, to them, safety is just another business risk, which comes after production, or profit, and approval from the various authorities, and even that can be gotten around with clever paperwork. To be successful with safety, a company needs a proper *safety culture*, which is something that comes from the top down. A good tool for this is an Occurrence Management System, for which Quality Assurance does nicely.

The whole subject can be divided into several broad areas, some of which have their own chapters in this book, because they have effects in their own right. They include (in no particular order):

- *Decision Making*, which is probably the most significant area for the single pilot, which depends on correct operation of the body, or

- *Physiological Factors*

- *Communication*, which can be affected by workload, fatigue, distraction, and other competing priorities, including anger

- *Workload Management*, also significant for the single pilot, which might include prioritisation, anticipation and the use of SOPs

- *Error Management*

- *Situational Awareness*

- *Commercal Pressure*

- *Stress*

- *Risk Management*

However, there will be a fair bit of crossover between them all, as some cannot be treated individually.

KEEPING COMPANY

As a single pilot, you will mostly be involved in charter or corporate work, especially when flying helicopters.

Charter

If scheduled flying is like bus driving, then charter flying is a taxi service, which means you are on call twenty-five hours a day with everything geared to an instant response to the customer, leaving you unable to plan very much. Don't get me wrong; this can be fun with plenty of variety and challenge in the flying, but the downside is an Ops Department that lets you do all the work yourself, and being left hanging around airports or muddy fields while your passengers are away (with missed meals, getting home late, etc.). Charter Flying is also where your other skills as Salesman and/or Diplomat come into play, as you will be very much involved with your passengers, who are more than just self-loading freight! Thus, while you can move relatively easily from Charter to Scheduled, it's not so straightforward the other way round. As a scheduled pilot, you rarely see your passengers, and the flying is very different. Charter (or Air Taxi) is intensive, single-handed and stressful work in the worst weather (you can't fly over it) in aircraft with the least accurate instruments. On a day flying charter, you could be working at almost any time, provided the Duty Hour limits are not exceeded. Departures are inevitably very early, as businessmen need to be where they're going at approximately the start of the working day and return at the end of it, so some days can be very long.

As you're only allowed a certain number of hours on duty, there's a continual race to minimise them, sometimes working like a one-armed paper-hanger to keep up with everything. The flight plan has to be filed, the weather checked (as well as the performance and the aircraft itself), the passengers' coffee and snacks must be prepared and they must be properly briefed and looked after (that's just the start). Usually, the only thing that can usefully be done the day before is to place the fuel on board, and even that can be difficult if the aircraft is away somewhere else. The flight itself is busy, too. As it's single-pilot, you do the flying, navigation and liaison with ATC. By contrast, the time at your destination is very quiet - after you've escorted your passengers through security and seen them safely on their way (the terminal's naturally miles away from the General Aviation park) you have to walk back to tidy up, supervise the refuelling, do the paperwork and have your own coffee (if there's any left) while preparing for the return journey.

If you're in a place you haven't been to before, you could always see the sights, but airfields are usually well away from anything interesting, with very few buses to get you there anyway. After a while, all you remember will be the same shops, so the general thing is to join the rest of the "airport ghosts", or other pilots in the same boat as you, and find a quiet corner to read a book. You may as well go to the terminal, because you have to meet your passengers there, but constant announcements could drive you out to the aircraft again (however, while you may be on time to meet them, your passengers will very rarely be on time to meet you!)

In Charter, it's also a luxury to have more than one day off in a row, and those you do get are needed by law, or turn up by surprise where you don't fly if business is bad, even though you've still gone into the office (the normal routine is - if you don't fly, you're not on duty, but common sense dictates that, if you're in the office doing something that is traceable, such as doing exams with a date on them, you'd better put down the hours). Some companies don't allow any leave at all during Summer, in the busy season, and only a week at a stretch if you do get it.

Corporate

Corporate flying, where you run the Flight Department for a private company, is similar to Charter, but not Commercial Air Transport, so the requirements (and paperwork!) are not so strict. Having said that, most corporate Flight Departments are run to Commercial standards, or better, and there is, naturally, no excuse for letting your own standards slip. One distinguishing feature is the way the Corporate world regulates itself - high performance intercontinental aircraft follow pretty much the same rules as single-engined General Aviation ones, and it's a credit to the people in it that things run so well.

In the Corporate world there are two types of Company. The first is the large conglomerate, where the aircraft is just as much a business tool as a typewriter is. You are genuinely a Company employee, people are used to the aircraft, you collect customers and move Company personnel around, from the Chairman to the workers, and your decisions as a professional are respected. There is a high degree of job satisfaction in this type of work, especially as you will build up relationships with regular passengers.

On the other hand, you might end up where the aircraft is the personal chariot of the Chairman, with you as its chauffeur (or, if you look at the books carefully, a gardener!), in which case nobody else gets to use it and what you think doesn't matter, because the sort of person who is dynamic enough to run a large company single-handed also thinks the weather will change just for him, and you're constantly under pressure to try and find the house in bad weather, which, naturally, hasn't got a navaid within miles. Unless you can establish a good personal relationship with your passenger, or have an extremely strong character, you are unlikely to get much job satisfaction here, especially if the company is family-run and you get to take the kids to horse shows, etc. at weekends.

Having said all that, there are some decisions that are not yours to take, whoever you work for. Unfortunately, you are only In Command where technical flying matters are concerned. If it's legal to fly then, strictly speaking, it's nothing to do with you whether it's sensible or not - it's an operational decision. If the Chairman (or Ops) wants you to fly and risk being left to walk if things get too bad, then it's entirely up to them - it's their money. For example, say you check the weather the night before and advise your passengers to go by car, because, while the destination and departure will be OK, the bit in the middle is iffy and there's no real way of knowing what it's like unless you go there and have a look (this is assuming a VFR flight in a helicopter, although the same principles apply elsewhere).

However, they must get there and the timings mean they can't delay things till the weather gets better, so it's the car or flying - a straight choice. If your man wants to try and fly, and risks missing the meeting at the other end because you refuse to either start or carry on when it becomes illegal or unsafe, then that, I suggest, is up to him. Please note that I'm not advocating flying in bad weather as a normal procedure! The problem is not just your ability to fly in those conditions, but what might happen later, such as 15 minutes afterwards, when you can't find your way back (or meet someone coming the other way!).

A major plus about Corporate Aviation is the way companies spend money on their flagship. It's a curious fact that, despite the higher standards that Commercial Air Transport demands, I have never yet seen a badly maintained Corporate aircraft and very few badly run Corporate Flight Departments, but decidedly the opposite has been the case in the commercial world. Corporate work sometimes pays the most, at least where smaller aircraft are concerned, but the jobs are less stable, as the aircraft is usually the first thing to go when the Company gets into financial difficulties. This often depends on how it is perceived by other parts of the organisation, so perhaps you could add marketing to your list of occupations.

Decisions, Decisions

A s mentioned in the introduction, your licence means the authorities consider that you have enough training to make decisions, but it cannot, and does not, cater for every situation.

Aviation is noticeable for its almost constant decision making. As you fly along, particularly in a helicopter, you're probably updating your next engine-off landing point every five seconds or so. Or maybe you're keeping an eye on your fuel and continually calculating your endurance in view of unexpected weather, or wondering whether to go or not go at V_1. It all adds to the many tasks you're meant to keep up to date with, because situations are always changing.

In fact, a decision not to make a decision (or await developments) is also a decision, always being aware that we don't want indecision. To drive a car 1 mile, you must process 12,000 pieces of information - 200 per second at 60 mph! It has to be worse with flying, and possibly over our limits - we can begin to see that our capability for processing information (below) is actually quite marginal, and is vulnerable to fatigue and stress - the most demands are at the beginning and end of a flight, but the latter point is when you are most tired (in fact, your heart rate is most just after landing).

A decision is supposed to be the end result of a chain of events involving judgment, after which you choose between alternatives. The process involves not only our eyes and ears which gather data, but our attention, which should not be preoccupied all the time. The human body is not a multi-tasker, and to keep track of what's going on it's necessary to split your attention for a short period between everything, typically a split second at a time.

Although decision making is a process involving several steps, things always seem to happen at once, so it's important not to get fixated on one thing at the expense of another, which is typically what happens when flying in bad weather. Gather all the information you can in the time available or, better still, get in the habit of updating information you're likely to need in an emergency as the flight progresses. Better yet, have a plan but, in normal life, what usually happens with a decision is that the thinking comes afterwards. In other words, there is an emotional element. If you go shopping for a house, for example, you might look at the outside and decide you like it there and then, until you discover that there is a factory around the corner that works all night, or the shops are too far away to walk to, or the neighbours are nasty. We must learn to do it the other way round, or at least reduce the emotional content!

STEPS INVOLVED

Normally, the need for a decision is triggered by recognising that something has changed, or that an expected event did not happen as it should. Thus, *situational awareness* is a fundamental part of the process, which is basically:

- Taking in information
- Trying to make sense of it
- Carrying out some action as a result

The point about decision making, as distinct from problem solving, is that the possible solutions are already known - you are faced with various alternatives, from which you have to make a choice. Problem solving involves reconciling a present position with a goal, with no obvious way of getting there - it is an attempt to achieve the goal through a series of logical stages, which might include *defining the problem*, *generating possible solutions* and *evaluating* them, which finally leads to the decision making process - actually, the first three options out of the steps below which are officially involved with making a decision:

- **G**ather all relevant information, using your senses
- **R**eview it
- **A**nalyze alternatives, keeping situational awareness (the big picture)
- **D**ecide and Do - make your choice and act on it
- **E**valuate the outcome - and be prepared to start all over again

You will notice that the problem solving comes first and the decision making comes late in the process.

Problem solving has two types of thinking associated with it. *Convergent* thinking brings information together, and *divergent* thinking generates different answers to one problem. The former requires more initial effort, whereas the latter requires more work towards the end.

The above decision making steps are not rigid, but may be merged or even repeated in a situation. For example, when adverse wether is ahead, you might get the updated weather, then vary the route or land to wait it out. Then you might get airborne and find you have to do it all over again, but this time land for refuelling, before getting airborne once more. The whole thing can be a continuously evolving process, which can be made quicker if some experience has already been gained, hence the value of training, which can allow you to make short cuts, if the situations are similar.

You gather information through the senses, but these don't always tell the truth, which we will look at later (of course, the information itself may be suspect). Thus, you recognise a change, assess alternative actions, make a decision and monitor the results. This can be enhanced with awareness of undesirable attitudes, learning to find relevant information, and *motivation* to act in a timely fashion.

The choices we make are affected by a complex range of factors, such as emotions, social context and uncertainty. Two circumstances where past experience can hinder decision making include *mental set* (or *rigidity*), where an older solution is still used, even when more efficient ones exist (which could be called *reproductive thinking*, rather than *productive thinking*), and *functional fixedness*, where we fail to see other solutions than the normal ones (in other words, think out of the box). *Confirmation bias* is the tendency to search for information to confirm a theory, while overlooking contradictory information, and can be likened to making the ground fit the map, rather than accepting the fact that we are lost.

Other influences include:

- *Belief Perseverance* - the tendency to cling to a belief, even if the evidence suggests otherwise

- *Entrapment* - when you've gone too far to withdraw from a situation, say, due to the costs involved

- *Overconfidence* - overestimating the accuracy of current knowledge

- *Expectancy* - or *Perceptual Set*, which can affect the perception of the world and what you do with the information. That is, your brain constructs its model of the world according to what it should be, and not what it is

- *Framing* - or the way information is presented (a 50% success rate is the same as a 50% failure rate)

INFORMATION PROCESSING

"Most people are woefully inadequate processors of information, who stumble along ill-chosen paths to reach bad conclusions"

In flight, you take on the role of an information processor - in this, you have a unique talent, in that a decision can be made without having all the relevant information to hand. If you were to ask a computer to choose between a clock that was gaining five minutes a day, and one that had stopped completely, it would probably choose the one that had stopped, because it was accurate twice a day, as opposed to once every 60 days or so. The point is that machines cannot discriminate, and they need *all* relevant information, which is good if you just want them to report facts, as with instruments, but not for making decisions.

Physical stimuli, such as sound and sight, are received and interpreted by the brain. *Perception* at this point means converting that information into something meaningful, or realising that it's relevant to what you're doing. What comes out depends on past experience of those events, your expectations, and whether you're able to cope with the information at that time (or are even paying attention to the situation). Good examples are radio transmissions from ATC, which you can understand, even if you can't hear them properly, because you expect certain items to be included, and you know from experience that they're bad anyway. The danger, of course, is that you may hear what you want to hear and not what is actually sent! (see *Communication*). As mentioned above, the human body is not a good multi-tasker, and to keep the various balls in the air over a typical flight, we must learn to *prioritise* and shift through tasks rapidly, which depends on how much attention the primary task is demanding. This can be reduced by using standard procedures, as with R/T and SOPs - the less thought secondary tasks require, the less attention they take up, especially when an external event happens to upset those well-made plans and flood the system.

Each decision you make eliminates the choice of another so, once you make a poor one, a chain of them usually follows. In fact, a decision-making chain can often be traced back up to and over fifty years, depending on whether the original cause was a design flaw (the F-15 and F-16, for example, are functionally identical, except that the speed bands go the opposite way in each aircraft. What bright spark thought that one up?) Another factor is the data itself; if it's incomplete, or altered through some emotional process, you can't base a proper decision on it. So:

- Don't make a decision unless you have to (keeps choices open)

- Keep it under review once you've made it

- No decision can be a decision (but watch for indecision)

Most important, though, is to be prepared to *change* a decision! (the Captain in the Dryden Accident should not have tried to take off a second time). Of course, the nature of most incidents means there's no time for proper evaluation, and you have to use instinct, experience or training.

There are two decision-making processes that affect us, both of which really speak for themselves - *ample-time* and *time-critical*.

Ample-Time Decision Making

You start with the awareness of a situation, which means having some idea of the big picture (similar to the continual updating mentioned above). *Situational awareness* here refers to your awareness of all relevant information, past or present, conscious or subconscious, which includes your cultural background (and given all that, it's no wonder people react to situations differently). Of course, you have to know how things *should* be to recognise what's wrong! You need *vigilance* and *continual alertness*, with regard to what *may* happen on top of what *is* happening, which is difficult at the end of a long day. Being a pilot, most of the information you will base a decision on comes from your instruments and navigation equipment, but this can be affected by your physical state, discussed in *Physiological Factors*.

There are three elements to the evaluation process. *Diagnosis* comes first (which is more of a skill than is thought), followed by the *generating of possible solutions* and the *assessment of any risks*, further described below.

When evaluating a situation, stay as cool as possible and not let emotions cloud your decision - that is, do not let false hopes affect your thinking.

Time-Critical Decision Making

This is where decisions have to be made quickly, based on past experience or training, often with no time to be creative or think up new solutions. In other words, time dictates your decision, and this is where checklists and SOPs can help, because they will be based on other peoples' experience (training is supposed to make your actions as near to reflex actions as possible, to make way for creative thought).

Drills, as per the Ops Manual, and checklists do the same thing on a different scale. Their purpose is to provide a framework on which to base good decision-making, as well as making sure you don't forget anything. SOPs are there to provide standardisation in situations where groups are formed and dissolve with great regularity, such as flight crews, as supported by checklists and briefings.

Although a checklist doesn't contain policy, it does at least stimulate activity, since the first response of most people in an emergency is to suffer acute brainfade. Either that, or you shoot from the hip, which is equally wrong. Checklists and drills are in the Company's Ops Manual and are intended to be followed to the letter (they are not always based on the Flight Manual drills, which must be followed to comply with the C of A). Whilst they have their uses, though, they can't cater for every situation, and you may have to think once in a while. In such circumstances, it pays to have prehandled many emergencies (i.e. updating landing sites as above), but, otherwise, actions take place in two modes, the *conscious* and the *automatic*. The former can be slow and error-prone, but has more potential for being correct. The latter is largely unconscious and therefore automatic, but it only relies on a vast database of information (or experience), and is not creative of itself - a problem that may affect inexperienced pilots.

MAKING PLANS

Where time is critical, such as whether to stop or go at V_1, it pays to have a plan ready in case something goes wrong, which is where a preflight briefing comes in (run it through your head by yourself if there's nobody else there). However, there's no point in having a plan if you don't *execute* it! Many accidents are caused because the original plan wasn't followed.

Memory

Memory is a feature in human information processing. We need it to learn new things - without it, we could not capture information, or draw on past experience to apply it in new situations (i.e. remembering). Thus, there are three processes involved in using memory, *input* (or *encoding*), *storage* and *retrieval*, any one of which can fail and make you think you're losing your memory, though this can depend on whether the items are placed in *short term* or *long term* memory (see below). However, to encode something in the first place, it must be given *attention*, which ultimately depends on whether it can be *perceived* against all the other stuff going on (discussed elsewhere).

Most psychologists (by no means all!) agree there are 3 types of memory:

- *Instinct*, what Jung called "race memory", which gives an immediate (gut reaction) response to a stimulus, like being hard-wired. Some psychologists call this *sensory memory*, as it provides a raw reaction to sensory input, and it can retain information long enough to allow you to decide whether a stimulus is important or not. The perception of a lighted cigarette moved across a darkened room being seen as a streak of light as opposed to a series of dots would indicate the existence of sensory memory, and your ears behave the same way, too. Information that does not get lost from sensory memory gets passed on to..........

- *Short Term*, or *Working*, Memory, which is for data that is used and forgotten almost instantly (actually, nothing is ever forgotten, as any psychologist will tell you, but the point is that Short Term Memory is for "on the spot" work, such as fuel calculations or ATC clearances, and figures greatly with situational awareness, which can follow short term memory's limitations). It can only handle somewhere between 5-9 items at a time (that is, 7 ± 2), unless some tricks are used, such as grouping or association (*chunking*), meaning that what can be held in short term memory depends on the rules used for its organisation (mnemonics are also good), which are actually kept in long-term memory, mentioned below. This is probably what Einstein was referring to when he thought that, as soon as one fact was absorbed, one was discarded. Data in short term memory typically lasts about 10-20 seconds, and is affected by *distraction* - there are only 27 lines to the Xanadu poem, because Coleridge was disturbed by the milkman. Because the capacity of short-term memory is so limited, items must clamour for attention, which may be based on *emotion*, *personal interest*, or the *unusual*. As mentioned, you can extend working memory's capabilities, either by *rehearsal* (mental repetition), or *chunking*, which means associating it with something meaningful - you could break up the information into sequences, as you might with a telephone number. The sequence of letters ZNEBSEDECREM becomes a whole lot easier to remember once you realise it is MERCEDES BENZ backwards, and suddenly your short term memory has 5 or so spaces left for more information (you can have 7 ± 2 chunks). However, expertise can also increase short term memory capacity.

Just to prove that short term memory really is limited, read out the following 15 words to a few people, taking one or two seconds per word, and get them to write down as many of them afterwards as they can remember. Most people will get 7. What is more interesting, however, is that some (around 55%) will include the word *sleep*, even though it wasn't there in the first place, which is proof that we sometimes hear what we want to hear, and that eyewitness testimony can be suspect, which is why the test was developed in the first place (by Washington University in St Louis).

The words are: *bed, rest, awake, tired, dream, snooze, wake, blanket, doze, slumber, snore, nap, peace, yawn, drowsy.*

In summary, working memory can handle speech-based and spatial information, has limited capacity, and cannot retain data for more than a few seconds without effort and resources being expended. It is also subject to interference.

Ultra short term memory has a duration of about 2 seconds, and acts like a buffer, that it, it stores information until we are ready to deal with it, although there are suggestions that this is actually handled by control processes such as rehearsal, or repetition.

Unfortunately, you cannot do any chunking or other type of association without activating..........

- *Long Term Memory,* where all our basic knowledge (e.g. memories of childhood, training, etc.) is kept - you might liken it to the unconscious (Chapter 5), with more capacity and ability to retain information than short-term memory, although we are only aware of the information in it when it surfaces in working memory. Where training is concerned, many processes can be carried out automatically, with little thinking. Repetition (or *rehearsing*) is used to get information into it, combined with organising it, placing it into some sort of context or associating it with an emotion (when studying, concentrate on the *meaning* rather than the subject matter). It is subdivided into *semantic memory,* based on things learnt through rule-based learning, and *episodic,* from specific events, for knowledge-based behaviour.

The reason why long term memory is required for association purposes is because it contains the rules used to give the items meaning. For example, chess players can have excellent short term memory for positioning pieces on the board, *if the rules are obeyed.* Upon random positioning, short term recall reverts to normality. People with brain damage (after accidents, etc) can often remember only one type of information, which supports the idea that the above types of memory are quite distinct, and that information can go directly into long term memory.

RESPONSES

Following a decision, based on a stimulus, there is a response. However, one resulting from excessive pressure is more likely to be based on insufficient data and be wrong than a more considered one, assuming time permits. Don't change a plan unnecessarily; a previously made one based on sound thinking is more likely to work than one cooked up on the spur of the moment, provided, of course that the situation is the same or similar. A correct, rather than rapid reaction is appropriate. Like revenge, decisions are best when cold (and dished out in instalments!) - those made the night before are likely to be better than those made during a flight, if the situation has not changed.

Response times will vary according to the complexity of the problem, or the element of expectation and hence preparedness (we are trained to expect engine failures, for example, but not locked controls, so the reaction time to the former will be less). Pushing a button as a response to a light illuminating will take about 1/5 th of a second, but add another light and button and this will increase to a second or so. An unexpected stimulus increases reaction time to nearly 5 seconds.

There is a time delay between perceiving information and responding to it, which is typically 3.4 seconds. The reason we don't take this long to answer people in normal conversation is because we are anticipating what they are going to say, which could lead to misinterpretation if you don't have body language to help, as with using radio.

ERRORS

An error, officially, arises when a planned sequence of activities fails to achieve the intended outcome, where random external intervention is not involved. *Latent* errors have consequences that lie dormant, while *Active* errors have consequences that are felt almost immediately. In fact, human error can be present at four levels:

- *unsafe acts* (errors & violations)

- *predispositions to unsafe acts*

- *unsafe or inadequate supervision*

- *organisational influences*

One working definition of human error is *"where planned sequences of mental or physical activity fail to achive intended outcomes, not attributable to chance."*

Error Management

New pilots naturally make mistakes during training - experienced pilots tend to have monitoring errors, and are more likely to think they are flying an older type. There are three ways of coping with errors:

- Avoiding them in the first place (that is, not getting into a position that requires your superior skills to get out of)

- If they happen, detecting and trapping errors before they become significant

- Sorting out the mess afterwards

The *Zero Defect Program* tries to eradicate errors by encouraging very high levels of motivation with rigid training and checking, although it ignores the influence of external errors. The *Error Cause Removal Program* tries to anticipate them. There have also been attempts to remove the human from the system altogether (although someone still has to program the computer!). However, whoever is in charge has finally combined reducing the causes of errors, with reducing their *consequences*. That is, as mentioned before, they have finally realised that sh*t happens, and have tried to make clearing up the mess easier. For more details, read Professor James Reason's book, *Human Error*, in which he points out that the sequence of human events in an accident can be likened to several slices of Swiss cheese (the stuff with holes in), with the holes as opportunities for accidents. On the day that the holes line up, something will happen.

The idea of having a non-punitive approach to errors is to encourage people to report them, so that others don't repeat them. There's no point, for example, in introducing penalties in to a reporting system (so that if you report yourself, you get punished!), because no errors will be reported. All it will do is make the Safety Officer look good!

Situational awareness, or being aware of what's going on, is your biggest weapon against error. Pilots who read newspapers on the flight deck are behind the aircraft, as are those who devote too much attention to detail.

Internal Influences

SENSING ERRORS

Errors must be detected for them to be reacted to - a good example is airspeed, which is often inferred from the surroundings, as when flying in mountains.

PERCEPTUAL ERRORS

These arise when interpretation is faulty, as influenced by context, data and expectancy.

ACTION SLIPS

After the wrong sequence of actions (raising the gear instead of flaps).

DECISION MAKING

Undue weight may be given to one factor, leading to a wrong conclusion.

FALSE HYPOTHESIS

One example arises from the Air Transat Airbus that ran out of fuel over the Atlantic, where the computer indications of low fuel were ignored, because they were not credible.

DISTRACTION

This can lead to false assumptions being made - the entire crew of the 1011 that crashed into the Everglades assumed the aircraft was under control while they tried to deal with a landing light problem.

MOTIVATION AND AROUSAL

Unmotivated or under-aroused people may commit more errors.

External Influences

THE SHELL MODEL

The individual letters of the word *SHELL* stand for *Software*, *Hardware*, *Environment* and *Liveware*, which are representative of the influences on the typical pilot. Hardware, naturally enough, is the mechanical environment, Environment covers such things as hypoxia, temperature, etc., whilst Liveware copes with interactions between the pilot, checklists, resources, other people, including the pilot himself.

Liveware-Hardware

The interface between liveware and hardware is still a source of errors, though perhaps not as much as in the early days. Where the hardware is poorly matched to the human, errors can and will occur - the 3-needle altimeter, for example, was a classic example of poor design that led to accidents. EFIS displays also fail to show patterns and trends, however good they might be at displaying maxima and minima.

Liveware-Software

Liveware-software problems can occur when checklists and manuals are poorly written or indexed.

Liveware-Environment

Physical and psychological stress significantly increases the probability of errors. Noise, vibration, temperature and heat all need to be carefully controlled, as do work patterns and shifts. A poor working environment will affect motivation, which itself will increase the likelihood of errors.

Liveware-Liveware

Deficiencies in teamwork and crew cooperation can have disastrous results, hence the need for books like this.

Physical

These are the influences that your body is subjected to.

THE ENVIRONMENT

Conditions under which an aircraft is operated. You may be remote, in a busy area, or just cold. You can feel cooler because moisture is evaporating from your skin at an advanced rate in dry air - humidity needs to be 60% at 18°C for comfort (the body operates comfortably between 18-24°C).

TIME

Pressure from customers and employers to keep to deadlines.

AIR QUALITY

Not only can haze or mist reduce visibility, it can be irritating, or smelly, or deadly. CO (carbon monoxide) is a colourless, odourless gas which has a half-life of about 6 hours at sea level pressures, so a quarter is present after 12 hours. It typically gets into the cockpit from faulty exhausts, but also comes about where something is burning without an adequate air supply, or where combustion is incomplete. One characteristic symptom of CO poisoning is cherry red lips (use 100% oxygen to recover).

FLICKER

This occurs when light is interrupted by propeller or rotor blades (see *Flicker Vertigo*, below). It can cause anything from mild discomfort to fatigue, and even convulsions or unconsciousness. Flicker certainly modifies certain neuro-physiological processes - 3-30 a second appears to be a critical range, while 6-8 will diminish your depth perception (the Germans set their searchlights to flicker, during WWII, to get up the nose of bomber pilots). Hangovers make you particularly susceptible.

NOISE, VIBRATION & TURBULENCE

Prolonged amounts of any of these is fatiguing and annoying - noise is particularly prevalent in helicopters, especially with the doors off, and vibration at the right frequency (8-12 Hz) causes back pain, as anyone who has flown a Bell 206 will tell you (see *Whole-Body Vibration*). Others include:

- 1-4 Hz - Affects breathing

- 4-10 Hz - Chest and abdominal pain

- 10-20 Hz - Headaches, eye strain, throat pain, speech disturbance & muscular tension

ERGONOMICS

Under this comes cockpit design and automated systems. Here's an illustration of how bad design can be the start of an *event chain*:

A relatively inexperienced RAF Phantom (F4) pilot had a complete electrics failure, as if being over the North Sea at night in winter wasn't stressful enough. For whatever reason, he needed to operate the Ram Air Turbine, but he deployed the flap instead, as the levers were close together.

Of course, doing that at 420 knots made the flaps fall off the back, and the hydraulic fluid followed. Mucking around with the generators got the lights back on, and he headed for RAF Coningsby, with no brakes. Unfortunately, the hook bounced over the top of the arrester wire, so he used full afterburner to go around in a strong crosswind, but headed towards the grass instead. The pilot and navigator both ejected, leaving the machine to accelerate through 200 knots, across the airfield at ground level.

Meanwhile, the Station Commander was giving a dinner party for the local mayor in the Mess, and the guests had just come out on the steps (near the runway), in time to watch the Phantom come past on the afterburner, with two ejections. The mayor's wife was just thanking him for the firework display when it went through a ditch, lost its undercarriage and fell to bits in a field.

The Fire Section had by this time sent three (brand new) appliances after it without any hope of catching up, but they tried anyway. The first one wrote itself off in a ditch because it was going too fast, the driver of the second suddenly put the brakes on because he realised there had been an ejection and that he might run over a pilot on the runway, at which point the number three appliance smashed into the back of him.

We are in a similar situation - how many times have you jumped into the cockpit of a different machine, to find the switches in a totally different place? This doesn't help you if you rely on previous experience to find what you need (in emergencies you tend to fall back to previous training), so the trick is to know what you need at all times, and take the time to find out where it is (read the switches, like you have to in the AS 350).

Another example of design being a factor in an accident is what happened to a V22 (US Marine tilt totor), which had the power controls changed to resemble those of the Harrier, because the Colonel in charge used to fly them. When another pilot with a helicopter background used the machine, he used the control in the opposite sense, got more power when he was expecting less, having lowered the lever, and made a situation worse, which led to a crash.

Physiological (The Body)

Refer to the *Physiological Factors* chapter.

Psychological

Refer to the *Psychological Factors* chapter.

Organisational Factors

That is, the people we work for, or with. An organisation is *a structure within which people work together in an organised and coordinated way to achieve certain goals.* The culture of the organisation can have a significant bearing on how people perform within it. Their goals may conflict, for example, resources also may be insufficient, as may planning or supervision. We all know about pressures, commercial or otherwise.

Automation

A lot of work is done for you by machinery or computers, which are just electronic machines. On the one hand, it is a good thing, because it can take much of the routine work away from you, and flight management systems can operate an aircraft very fuel-efficiently. For example, the FADEC on the Bell 407 incorporates many monitoring functions. On the other hand, it can induce a feeling of *automation complacency* (too much reliance on the machine) and lead you not to check things as often as you should (*reduced vigilance*), or push the envelope, as when using a GPS in bad weather - with much of the navigation task taken away from you, it is very tempting to fly in worse weather than you can really cope with (flying in bad weather is like sex - the further you get into it, the harder it is to stop).

However, a major benefit is the integration of many sources of information and its presentation in a clear and concise manner, as with the glass cockpit, and providing a major contribution towards situational awareness, provided the pilot keeps a mental plot going, as the information presented can be highly filtered.

Cultural Factors

What's normal for one person isn't for others - in many countries, red is an extremely unlucky colour. Even within our own trade, there are fixed wing and rotary pilots, military versus civilian, jet against piston, etc., each with their own ways of thinking.

RISK MANAGEMENT

A skilled pilot who takes risks is a bigger problem than an average one who is prudent and cautious

"Carelessness and overconfidence are often more dangerous than deliberately accepted risks"

Wilbur Wright

Uncertainty about a situation can often indicate risk. One definition of it is the chance that a situation, or the consequences of one, will be hazardous enough to cause harm, injury or loss. Another is that a risk arises every time a person is in the presence of a hazard. To have absolutely no risk, of course, we shouldn't take off at all, but that's not what we're here for, so we have to have some method of evaluating risk against a yardstick to get the job done, or balance profitability with safety. *Risk management* is the key, best used in an ample-time decision-making situation, where time is not critical. For example, in a helicopter, it can be more dangerous to avoid the height/velocity curve (say when coming out of a confined area) than to be in it for a few seconds. Part of the pilot's job is to decide which of the risks presents the least hazard - that is, is there a greater risk of colliding with something when coming out of the clearing than having an engine failure? In other words, Risk Management means measuring the degree of harm against that of exposure.

It is a decision-making tool that can be applied to either eliminate risk, or reduce it to an acceptable level, preferably before takeoff (things that stop you eliminating risk entirely would either be impracticality, or money). With it, you have to first identify a hazard, analyse any associated risks, make a decision and implement it (with a *risk strategy*) and monitor the results, with a view to changing things if need be. However, this depends on the *perception* of a risk, and the difference between yours, the management's and the customer's can be quite startling. Outside influences include weather, traffic and obstacles. Internal ones can be maintenance, fatigue, or the culture of the company.

Analysing Risk

There are two aspects to analyzing risk:

- Where is it?

- How significant is it?

The difference between perceived and actual risk depends on the amount of control you think you have, and familiarity. For example, it is a lot more risky to ride a bike through a busy city than it is to live near a nuclear power station, yet people still ride bikes and don't want to live near Sellafield. The former situation allows you more control (you can always get off and walk) and is more familiar.

Risk is equal to *probability* multiplied by the *consequences* of what you propose to do, and your *exposure*. You essentially have four choices - either not do the job, mitigate the effects of the risk, transfer it (buy insurance) or eat it (absorb the effects yourself). *High risk* means a high probabability of death, damage or injury, requiring appropriate procedures - possibly with none available at all and you have to think on your feet, which is where your training comes in. *Low risk* is a normal situation, where normal precautions are enough.

You could always try and prehandle situations - that is, make as many decisions as possible ahead of time, as part of your flight planning - most important, though is to leave yourself a way out. For example, always be aware, when dropping water, that you may have to get out of a hot hole with the load on - don't assume that the bucket will work and you will be light enough to escape! Is the weather closing in behind you? Have you gone into a confined area and boxed yourself in? Mountain pilots *always* have a way out - even after they've landed.

PHYSIOLOGICAL FACTORS

The human body is wonderful, but only up to a point. It has limitations that affect your ability to fly efficiently, as your senses don't always tell you the truth, which is why you need extensive training to fly on instruments - you have to unlearn so much. The classic example is the "leans", where you think you're performing a particular manoeuvre, but your instruments tell you otherwise. However, although the sensors in the eyes and ears may be sensitive, the brain isn't, and does not always notice their signals. Sometimes it even fills in bits by itself, according to various rules, which include your expectations and past experience. Thus, at each stage in the perception process, there is the possibility of error, because we are not necessarily sensing reality. For example, the reason why there is a *white balance* setting on a digital camera is because the brain interprets what is white in its own way and compensates all by itself - indoor bulbs actually glow quite red, and an overcast sky might have some blue in it, despite what you think you see. If the camera doesn't compensate, your pictures will be tinted.

THE SENSES

Officially, the five senses are vision, audition (hearing), olfaction (smell), tactition (touch) and gustation (taste). However, there is also *proprioception*, or the internal senses, which include kinaesthetic (muscular movement) and balance. Each has a sense organ which has receptors, to convert the specialist energy they detect into nerve impulses which are transmitted electronically to the relevant part of the brain.

The *absolute threshold* is the minimum level of stimulation (for a sensor) at which a stimulus is noticed. The increase in stimulation required for us to notice a change between two stimuli is the *difference threshold* or the *Just Noticeable Difference* (JND), which is otherwise known as *Weber's Law.* Changes between the two thresholds may not be noticed and may build up in flight to extreme attitudes, hence the need to watch those instruments.

After a stimulus is removed, particularly with sight, there might be an afterimage, which is caused by overstimulation of the receptors (light also has a momentum). There is more on perception in Chapter 3.

THE BODY

• •

Why do you need to learn about the body? Well, parts of it are used to get the information you need to make decisions with, and, of course, if it isn't working properly, you can't process the information or implement any action based on it. In the single-pilot case, it needs to be more efficient because there is nobody else to take over if you get incapacitated.

Body Mass Index (BMI)

This is calculated by your weight in kg divided by your height squared, in metres squared. If it is over 30, you are obese. This could lead to heart disease, and reduce your ability to cope with hypoxia, decompression sickness and G tolerance.

Central Nervous System

Whatever your body gets up to, the processes involved must be coordinated and integrated, which is the job of the central nervous system, with a little help from the endocrine system. Although coming down the ILS might seem to be automatic, the control responses that occur as a result of input from your eyes and ears, and experience, plus the feedback required from your limbs so that you don't overcontrol, are all transmitted over complex nerve cells (*neurons*) for processing inside the Central Nervous System, which consists of the brain and spinal cord, though, for exam purposes, it also includes the visual and aural systems (eyes and ears), proprioceptive system (internal senses) and other senses (see below).

Neurons don't touch each other directly - if a message needs to be transmitted, a chemical called a *neurotransmitter* (of which there are over 50 types) carries it across the small gap between them. Modern drugs pretend to be neurotransmitters, that is, they work by providing a "key" to the receptor's "lock".

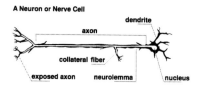

A Neuron or Nerve Cell

Peripheral Nervous System

This connects the Central Nervous System with the sense organs, muscles and glands, and therefore with the outside world. The PNS is divided into:

- the *somatic nervous system*
- the *autonomic nervous system,* which regulates vital functions over which you have no conscious control, like heartbeat and breathing (unless you're a high grade Tibetan monk, of course). It consists of:
    - the *sympathetic*
    - the *parasympathetic*

nervous systems. The former prepares you for fight-or-flight (see *Stress,* below) and tends to act on several organs at once, while the latter calms you down again, acting on one organ at a time. Being under the influence of fight-or-flight is like being in a powerful car in permanent high gear, which you can't do all the time - you need rest & relaxation to allow time for the parasympathetic system to kick in, such as meditation, or a snooze in the back of the helicopter. Being in such a high state of readiness all the time produces steroids, and can lead to depression.

THE BRAIN

Although the brain is only 2% of the body mass, it takes up to 20% of the volume of each heartbeat - its blood supply needs to be continuous, as it cannot store oxygen. Many of the brain's departments merge into each other, and work closely together, but it still has three distinct areas, namely the *central core,* the *limbic system* and the *cerebral hemispheres.*

The Central Core

This includes most of the brain stem, starting at the *medulla* where the spinal cord widens as it enters the skull. The medulla controls breathing and some reflexes that keep you upright. Also, the nerves coming from the spinal cord cross over here, so the right side of the brain is connected to the left side of the body, and *vice versa.* Slightly above the medulla is the *cerebellum,* which concerns itself with (smooth) coordination of movement.

The *thalamus* consists of two egg-shaped groups of nuclei. One acts as a relay station for messages, and the other regulates sleep and wakefulness. Just below that is the *hypothalamus,* which regulates endocrine activity (through the pituitary gland) and maintains normal body functions, in terms of temperature, heart rate and blood pressure, which are disturbed when under stress. For example, the body's core temperature should be between 35-38°C (normal is 37°C). It is maintained through mechanisms such as sweating, shivering, or goose pimples, when hot or cold.

Because of its role in responding to stress, and preparing the body for fight-or-flight, the hypothalamus is also known as the *stress centre*.

The Limbic System

This is closely connected to the hypothalamus. Part of it, the *hippocampus*, would apppear to have something to do with short-term memory, in that, when it's missing, people can remember things that happened long ago, but not anything recently.

Cerebral Hemispheres

These consist of the "grey matter" you see when looking at a picture of the brain (see left). Each half is basically symmetrical, but the left and right hemispheres are interconnected, with women having more connections between them than men, which accounts for their ability to think of several things at once (often contradictory!). Each hemisphere has four *lobes*.

The two hemispheres work in different ways, for two types of thinking:

- *Left Brain*, or logical. This governs language and is skilled in mathematics

- *Right Brain* - conceptual, for artistic thinking

Note that, although the two hemispheres work differently, they still work very much together.

THE ENDOCRINE SYSTEM

This controls individual cells, through *hormones*, which come from glands like the pituitary. Like neurotransmitters, hormones are only recognised by certain types of cell, although they act over longer distances.

THE EYES

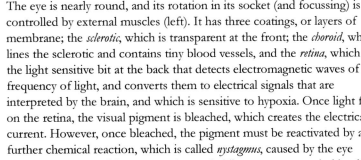

Vision is your primary (and most dependable) source of information - 70% of the data you process enters the visual channel. It gets harder with age to distinguish moving objects; between the ages of 40 and 65, this ability diminishes by up to 50%. However, this is only one limitation, and we need to examine the eye to see how you overcome them all.

The eye is nearly round, and its rotation in its socket (and focussing) is controlled by external muscles (left). It has three coatings, or layers of membrane; the *sclerotic*, which is transparent at the front; the *choroid*, which lines the sclerotic and contains tiny blood vessels, and the *retina*, which is the light sensitive bit at the back that detects electromagnetic waves of the frequency of light, and converts them to electrical signals that are interpreted by the brain, and which is sensitive to hypoxia. Once light falls on the retina, the visual pigment is bleached, which creates the electrical current. However, once bleached, the pigment must be reactivated by a further chemical reaction, which is called *nystagmus*, caused by the eye jerking to a new position, to remain steady. The movement period is edited out by the brain, and the multiple images are merged, so continuous vision is actually an illusion, as an *after image* is produced when light falls on the retina - that is, the image of what you are looking at remains there for a short period, as light has a momentum (try it by closing your eyes and looking at the picture that remains). As the eye does not need to be seeing constantly (and can therefore be regarded as a detector of *movement*), it can spend the spare time in repair and replacement of tissue. 30-40 images per second are taken in the average person, and an image takes about 1/50th of a second to register.

All this means you *see with the brain*, giving a difference between *seeing* and *perceiving* (it also means that vision problems can arise from the brain's processing ability and not the eyes themselves). This is because the eye's optical quality is actually very poor (you would get better results from a pinhole camera), hence the need for the brain, which can actually modify what you see, based on experience, and so is reliant on expectations. If the brain fills in the gaps wrongly, you get visual illusions. One classic illusion relevant to pilots is whiteout, which is defined by the American Meteorological Society as:

> *"An atmospheric optical phenomenon of the polar regions in which the observer appears to be engulfed in a uniformly white glow".*

That is, you can only see dark nearby objects - no shadows, horizon or clouds, and you lose depth perception. It occurs over unbroken snow cover beneath a uniformly overcast sky, when the light from both is about the same. Blowing snow doesn't help, and it's particularly a problem if the ground is rising. *Flat light* is a similar phenomenon, but comes from different causes, where light is diffused through water droplets suspended in the air, particularly when clouds are low.

The eye can, however, react quickly to changes in light, although it is slow to adapt from light to dark because a chemical (*visual purple*) needs to be created. The only part of the eye that sees perfectly clearly is in the centre of the retina, an area not much larger than a pinhead, called the *Macula Lutea*. Outside of that area, vision is blurred - if you look at the top part of this page, for example, you will not be able to see the rest clearly without shifting your vision. The illusion of seeing large areas clearly (that is, more than two words at a time) comes from the rapidity of shifting - attempting to do this otherwise means seeing without focussing, and results in eyestrain. Sometimes your eye and brain can get out of the habit of looking at one point together.

The *optic nerve* carries signals from the eye to the brain, the lens focuses light on to the retina, the *iris* controls the diameter of the pupil (the black bit in the middle where light gets through), and the *cornea* refracts light onto the lens. The area of sharp vision is actually very small (at 4 feet, the size of a small coin), because of the relatively small size of the *fovea*, in the central part of the retina. 5° away from the foveal axis, it reduces by a quarter, and one-twentieth at 20° away. You should be able to see another aircraft directly at 7 miles, or 2.5 miles if it was 45° off - at 60° it's down to half a mile! The reason why you must scan is because the eye needs to latch on to something, which is difficult to do with a clear blue sky. With an empty field of vision, your eyes will focus at short distances, about 1-2 metres ahead, and miss objects further away. The ratio of looking inside and outside the cockpit should be 5:15 seconds.

A high speed aircraft approaching head-on will grow the most in size very rapidly in the last moments, so it's possible for it to be hidden by a bug on the windscreen for a high proportion of its approach time (you might only see it in the last few seconds). Lack of relative movement makes an object more difficult to detect.

Peripheral vision also has a different neurological route to the brain. To prove this, stand up, cover one eye and make a fist with the other, to cover up the good eye's central vision. Now stand on one foot and do it again, with your forefinger and thumb making a small circle where the fist was - in other words, using your central vision to look through the circle. Standing on one foot is different each time!

Normal vision is described as 20/20, meaning that you can see at 20 ft what a normal person can see at 20 ft. If the ratio, as a fraction, is greater than 1/1, visual acuity is better than normal. Thus, 6/4 means you can see at 6 m what a normal person can only distinguish at 4 m. On the other hand, 6/9 is poor: Normal people can detect at 9 m what you cannot see above 6 m.

Clarity of vision is affected by:

- light available
- size and contours of objects
- distance of an object from the viewer
- contrast
- relative motion
- the clarity of the atmosphere

The field of view of each eye is about 120° left to right, and about 150° up and down. There is an overlap of 60° in the centre where binocular vision is possible, and which has a blind spot about 5° wide where the optic nerve leaves the eyeball and there are no rods nor cones (see below). The brain normally fills in the blanks.

Close your right eye, place the book about a foot from your eyes, and move it forward and back whilst looking at the helicopter on the right. The one on the left will disappear when its image hits your blind spot, so it's possible not to see an approaching aircraft through your windscreen if the conditions are right. The greatest visual acuity (to see small detail) is obtained by cones in the fovea, so you must look directly at an object to see it best. At night, look slightly to one side, as the rods that are sensitive to lower levels are outside the fovea, at the peripheral of the retina.

Depth Perception (useful when slinging) is the process of forming 3D images from 2D information, in our case, 2 sets, from our eyes - and it's all done in the brain. *Binocular clues* rely on both eyes working together in four main ways:

- *Retinal Disparity* - depends on the difference in images received by each eye, and gives an important clue to distance - look at your finger near and far

- *Stereopsis* combines two images into one 3D sensation

- *Accommodation* is a muscular clue to distance, from the change in curvature of the lens, which gets thicker as you focus on nearby objects and flattens with distant ones

- *Convergence* is another muscular clue where the eyes point more and more inward as an object gets closer (like on eye tests). That is, each eye sees an object from a different angle. By noticing this angle of convergence, the brain produces depth information over 6-20 feet. You judge speed by the rate of change of the angle of convergence

Monocular clues, used for longer distances, include *relative size* (larger objects are closer), *overlap* (an object covered by another is perceived as being further away), *relative height* (lower objects are closer), *texture gradient* (smoother surfaces are further away), *linear perspective* (more convergence means more distance), *shadowing, relative brightness* (nearer objects are brighter), *aerial haze* and *aerial perspective* (difference in focus and colour) and *motion parallax* (nearer objects move more). With reference to brightness, it has been shown that a pendulum swinging in a straight line in front of a person with one eye covered is actually seen as swinging in an ellipse.

The major causes of defective vision are:

- *Hypermetropia* - where the eyeball is too short, and images focus behind the retina (farsightedness). Requires a convex lens

- *Myopia* - where the eyeball is too long, and images focus in front of the retina (short sight). Needs a concave lens

- *Presbyopia* - the lens hardens, leading to *hypermetropia* and difficulty in focussing (comes with old age)

- *Cataracts* - the lens becomes opaque. Cataracts cannot be removed until they mature

- *Glaucoma* - pressure inside the eyeball that interferes with accommodation and blood flow to the retina

- *Astigmatism* - unequal curvature of the cornea or lens

To optimise night vision, spend some time getting used to lower levels of light (around 20 minutes), reduce the cockpit lighting to make instrument lights more visible and avoid getting blinded.

Rods and Cones

The retina is composed of ten very thin layers, with light sensors (actually, *neurons*) which are called *rods* and *cones*, in the ninth (their names arise from the way they are shaped). Each is more efficient than the other in different kinds of light. Rods are the more primitive sensors and respond in low light to varying shades of black. The output from three different types of cone (red, green and blue sensitive) is mixed to produce colour.

The point where the optic nerve joins the retina (on the nasal side, near the middle of the back) is mostly populated with cones, which work best in daylight and become less effective at night, or where oxygen levels are reduced (which is significant for smokers, whose blood has less oxygen carrying capacity), so you get a blind spot in the direct field of vision, which is why you see things more clearly at night if you look slightly to the side of what you need to see. The blind spot is normally ineffective because the brain superimposes two images. Rods cannot distinguish colours, either, which is why things at night seem to be in varying shades of grey (you see colours because the vibrations they give out are strong enough to wake the cones up). The brain mixes the colours received by the cones, and the most common colour blindness is red/green. Rods are sensitive to shorter wavelengths of light, so in very low light, blue objects are more likely to be seen than red (neither will be in colour), which is why cockpit lighting is sometimes red because it affects the rods used for night vision less than white light.

The *fovea centralis* is a small point on the retina where the first eight layers are missing, so the rods and cones are directly exposed to light (that is, the light doesn't have to battle through the first layers) for clearer vision at that point. The optical fibres from the right side of the right eye go to the right side of the brain, and those from the left side go to the left side of the brain. Ditto for the left eye, so each side of the brain has input from both eyes at once. The more your iris is open, the less *depth of field* you have, so in darkness it is difficult to see beyond or before a certain distance, and you may require glasses to help (the depth of field in photography is an area either side of the focus point in which everything is sharp. The wider the aperture, or iris, the shorter this distance is, and *vice versa*).

The retina contains enormous amounts of vitamin A, which is necessary for adapting between light and darkness. Too little vitamin A could therefore result in night blindness. The changeover from light to dark takes about 20-30 minutes and should always be allowed for when night flying (actually, the cones take 7 minutes, and the retina can take up to 45 minutes). The best way to use the eyes at night is to scan slowly, to allow offcentre viewing.

Night Myopia and Night Presbyopia

Night myopia (nearsightedness), also known as *twilight myopia*, causes some people who are slightly myopic in daylight to become more so after dark.

Presbyopia is a condition in which the crystalline lens of your eye loses its flexibility, which makes it difficult to focus on close objects. Also known as *red light presbyopia*, night presbyopia occurs in presbyopic individuals who are subjected to red light, which is found in some cockpits during night operations. Red light has the longest wavelength, so when you try to read instruments or charts in red light, the demand for accommodation is more than if you were using white light, making it difficult to read small print.

Space Myopia

This describes myopia experienced when there is nothing to look at outside the cockpit. For example, when flying VFR on top, clouds prevent you from seeing the ground, and the light they reflect reduces your visual cues. Your eyes will tend to lock-in on the instruments and remain fixated for that distance, so when you look outside, the resulting myopia could stop you seeing other aircraft. Look at the wingtips from time to time to allow relaxation of the ciliary muscles (the ones that control the shape of the lens for near and far vision).

Design Eye Reference Point

The DERP allows crews to obtain the best visibility outside and inside. The pilot compartment should be designed for a clear, undistorted, and adequate external field of vision, but not necessarily for what sizes of pilot it can hold, so seats need to be adjusted to position your eyes as close to the DERP as possible for the best views while manipulating the controls. You should be aware of the hazards and compromises associated with seating positions away from the DERP.

Refraction

The transparent part of the sclerotic is known as the *cornea*, behind which is the *lens*, whose purpose is to bend light rays inwards, so they focus on the retina. If this happens in front of it, short sightedness results (you get long sightedness with the point of focus behind the retina).

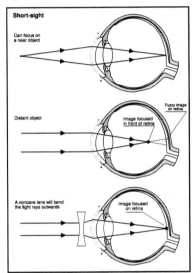
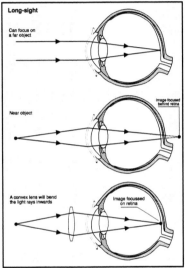

Both conditions cause blurred vision, correctable by glasses, that vary the refraction of the light waves until they focus in the proper place. Blurred vision can also be caused by stress, causing nervous tension, and excessive eye muscle activity leading to eyestrain. Just relaxing often helps this.

70% of light is refracted by the cornea, and 35% by the lens.

Optical Illusions

An illusion exists when what you sense does not match reality. One that pilots in the Persian Gulf get on hazy day is a sight of the rig they are going for 150 feet above them! You can't do much about illusions, as they are based on what your senses tell you, but you do need to be aware that they might happen and be prepared to ignore what you are being told.

Searching for an object in a swimming pool is difficult, because the light rays bend as they pass the surface and the object appears to be displaced. Similarly, rain on a windscreen at night gives the impression that objects are further away. A good fixed wing example of an optical illusion is a wider runway tending to make you think the ground is nearer than it actually is; a narrow runway delays your reactions, possibly leading to a late flare and early touchdown. Any object (like a runway) that you think is smaller than it actually is, will also appear to be nearer, and vice versa. If the object is brighter than its surroundings (a well-lit runway, or landing lights on an oil rig), you will think you are higher than you are, so on an approach, you might start early and be lower than you should. In haze, objects appear to be further away. Lack of ground features, as when landing over water, darkened areas and featureless terrain (as with snow) can give the illusion of being too high. An approach to a downsloping runway should be started higher, with a steeper angle, and one to an upsloping runway should be started lower, at a shallower angle.

When mountain flying, it's often difficult to fly straight and level because the sloping ground around affects your judgment. Similarly, you can't judge your height when landing on a peak. Even going to the cinema is an optical illusion; still frames are shown so quickly it looks as if movement is taking place - the switching is done in the brain. *Vectional illusions* are caused by movement, as when sitting in a railway carriage and wondering whether it's the train next to you or the one you're in that is moving. The *autokinetic effect* is the illusion that an object is moving, when it is actually your eye. Distant objects become less colourful and less distinct.

Distortion occurs when viewing objects through a windshield covered with rain, where water is thicker near the bottom (nearer to the windshield), causing a *prismatic effect* - like looking through a base-down prism, which tends to make objects look higher or closer than they actually are.

Helicopter pilots can experience the *waterfall effect* when hovering or in slow flight at low altitudes over water. The downwash causes the air to pick up water and to displace it upward at the edge of the blades and downward directly under them, so you might see drops of water going downward in your field of vision to give a climbing sensation. A corrective manoeuvre to descend will put the helicopter in the water.

THE EARS

These are important because an auditory stimulus is the one most often attended to. How many times do you answer the phone when you're busy, even though you've ignored everything else for hours?

Sound waves make the eardrum vibrate, and the vibrations are transmitted by a chain of linked bones known as the *hammer, anvil* and *stirrup* (collectively, the *ossicles*) to the *cochlea* in the inner ear, which is full of fluid. Thus, the eardrum is the boundary between the inner and outer ears.

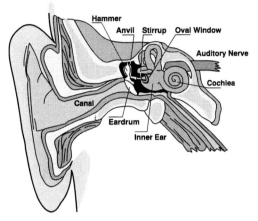

The cochlea is the coiled bit in the right hand side of the diagram below - it is a tube which narrows progressively. There are thousands of fibres of different lengths inside it which vibrate in sympathy at different frequencies. As some of the fibres get damaged (through too severe vibration), the ability to hear that frequency goes (they do not regenerate). *Presbycusis* is hearing loss with age, where the high tones go first. *Noise Induced Hearing Loss*, or NIHL, occurs through prolonged exposure to loud noise, usually 90 db and above. The fibres are linked to the brain and, as with sight, it is now, when the signal reaches the brain, that we "hear".

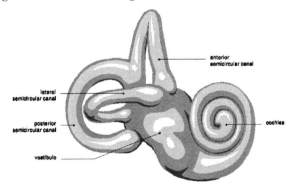

The *semicircular canals* are what we use to keep balanced (above). They are arranged at rightangles to each other, and use the fluid in the inner ear.

This acts against sensory hairs with chalky deposits on the end to send electrical signals to the brain so you can tell which way up you are (the chalky deposits are affected by gravity and detect *angular acceleration*). The *leans* happen because your semicircular canals get used to a particular sustained motion in a very short time. If you start a turn and keep it going, your canals will think this is normal, because they lag, or are slow to respond. When you straighten up, they will try to tell you you're turning, where you're actually flying straight and level. Your natural inclination is to obey your senses, but your instruments are there as a cross-reference. In fact, the whole point of instrument training is to overcome your dependence on your senses. Particularly dangerous is recovering from a spin of 2-3 turns, where you think you are turning the opposite way and enter another spin when you try to correct it. Eventually an extreme nose-up condition results, which turns into an extreme nose-down attitude and a tight graveyard spiral before entering Terrain Impact Mode.

The coriolis illusion is easily demonstrated with a revolving chair - sit in one, and get someone to spin it while you have your chin on your breast. When you raise your head sharply, you will find yourself on the floor inside two seconds. This has obvious parallels with flying, so make all your head movements as gently as possible, especially when making turns in IMC (mention of fluid, above, implies that if you are dehydrated, you may also get spatial disorientation - if you feel thirsty, you are probably already 5% there). You can get problems from colds, etc. as well, particularly a spinning sensation caused by a sudden difference in pressure between the inner portions of each ear.

The *Eustachian Tubes* are canals that connect the throat with the middle ear; their purpose is to equalise air pressure. When you swallow, the tubes open, allowing air to enter, which is why swallowing helps to clear the ears when changing altitude. Blocked Eustachian tubes can be responsible for split eardrums, due to the inability to equalise pressure. Since the eardrum takes around 6 weeks to heal, the best solution is not to go flying with a cold, but commercial pressures don't always allow this. If you have to, make sure you use a decongestant with no side effects. The audible range of the human ear is 20 Hz to 20 KHz, with the most sensitive range between 750-3000 Hz.

Blocked Sinuses

Although associated with the nose, the sinuses are actually hollow spaces or cavities inside the head surrounding the base of the nose and the eye sockets. Amongst other things, they act as sound boxes for the voice.

Being hollow, they provide structural strength whilst keeping the head light; there are normally between 15-20. Blockages arise from fluid that can't escape through the narrow passages - pain results from fluid pressure. Blocked sinuses can give you severe headaches.

Deafness

This can arise from many causes; in aviation, high-tone deafness from sustained exposure to jet engines is very common. Hearing actually depends on the proper working of the eighth cranial nerve, which carries signals from the inner ear to the brain. Obviously, if this gets damaged, deafness results. The nerve doesn't have to be severed, though; deterioration will occur if you don't get enough Vitamin B-Complex (deafness is a symptom of beriberi or pellagra, for example, from Vitamin B deficiency). You can recover from some deafness, such as that caused by illness, but not that caused by damage to the fibres in the fluid.

Disorientation

This refers to a loss of your bearings in relation to position or movement. The "leans" is the classic case, already mentioned above. To combat them, close your eyes and shake your head vigorously from side to side for a couple of seconds, which will topple the semi-circular canals. Motion sickness usually happens because of a mismatch between sight, feel and the semicircular canals, giving unfamiliar real or apparent motion (e.g. the leans). Medication can have unwelcome side effects, particularly on performance, which are normally not acceptable for flight crews.

During acceleration, it's possible to get the impression of pitching up (*somatogravic illusion*), making you want to push the nose down. It's more pronounced at night, when going into a black hole from a well-lit area, as confirmed by the artificial horizon, which suffers from the same effect. You get a pitch-down illusion from deceleration. The danger here is that lowering the gear or flaps causes the machine to slow down, which makes you think you are pitching down and want to bring the nose up, which could cause a stall at the wrong moment on approach. A good preventive measure, aside from trusting your instruments, is to keep hydrated.

A steady light flickering at around 4-20 Hz can produce unpleasant and dangerous reactions, including nausea, vertigo, convulsions or unconsciousness, which are possibly worse when you are fatigued, frustrated, or in a state of mild hypoxia. Military helicopter pilots are tested for *Flicker Vertigo* during selection, as the Sun flashing through the rotor blades can be a real problem.

Motion Sickness is one reaction to confusing stimuli, and air sickness can also occur when the body (actually the skull), is vibrated at frequencies less than 0.5Hz, which is common in turbulence. Keeping the head still and closing the eyes helps, but this is difficult when you are the PF!

THE RESPIRATORY SYSTEM

This consists of the lungs, oronasal passage, pharynx, larynx, trachea, bronchi, bronchioles and alveoli:

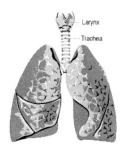
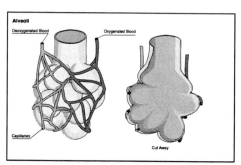

Air is drawn into the lungs, from where oxygen is diffused into the *haemoglobin* in the blood under pressure, which carries it to the tissues of the body, especially the brain, the most sensitive to lack of it. Blood is pumped around by the heart. Waste products in the form of carbon dioxide go the other way, via plasma to the lungs - it is the carbon dioxide level in the blood that regulates respiration, which is monitored by several chemical receptors in the brain that are very sensitive to it. Breathing is controlled by the autonomic nervous system, but can change according to your activity. Breathe with your stomach and chest - not only does this fill the lungs better, it stimulates the liver and heart. The normal rate of breathing is around 18 times a minute, exchanging .35-.65 litres of air.

The diffusion of oxygen into the blood depends on *partial pressure* (that is, its pressure in proportion to its presence in the mix - it follows *Dalton's Law*), so, as this falls, oxygen assimilation is impaired (although the air gets thinner, the ratio of gases remains the same). Even if you increase the proportion of oxygen to 100% as you climb, there is an altitude (around 33,700 feet) where the pressure is so low that the partial pressure is actually less than that at sea level, so just having oxygen is not enough, because, as altitude increases, the partial pressure of water vapour and carbon dioxide in the lungs also remains the same, reducing the partial pressure of oxygen in the lungs still further.

From 0-10,000 ft you can survive on normal air; above this, an increasing amount of oxygen relative to the other ingredients is required, up to 33,700 feet, at which point you require pure oxygen to survive (breathing 100% oxygen at that height is the same as breathing air at sea level. At 40,000' it is the equivalent of breathing air at 10,000 feet). Above 40,000 the oxygen needs pressure, meaning that you must exhale by force (also, exposure to O_3 becomes significant). Having said all that, your learning ability can be compromised as low as 6000 feet (*Source*: RAF).

Oxygen

Around the Earth is a collection of gases, called the *atmosphere*. 21% of it, luckily for us, is oxygen, but 78% is nitrogen, with 1% of odds and ends, like argon (0.9%) and CO_2 (0.03%), that need not concern us here, plus bits of dust and the odd pollutant. The nitrogen, as an inert gas, keeps the proportion of oxygen down, since it is actually quite corrosive. The water vapour content on average is around 1%, but can get as high as 4%.

Aside from keeping us alive by supporting combustion, oxygen has other benefits for the human body - for example, it has been noticed that, in the USA at least, only 1 in 7 athletes tend to get cancer, against a general level in the rest of the population of 1 in 4. Exhalation also provides a vacuum effect to suck lymph through the body, and with it, waste products from the cells (the lymph system does not have a pump like the heart).

The atmosphere is split into four concentric gaseous areas. Starting from the bottom, these are the *troposphere, stratosphere, mesosphere* and *thermosphere*, although the last two are not important right now. The first two are important, however, and the boundary between them is called the *tropopause*, a freezing layer of dry air, where any clouds are made of ice crystals. So, underneath the tropopause is the troposphere, and above it is the stratosphere, where the temperature remains relatively constant with height. The troposphere contains more than 80% of the mass of the atmosphere. Although the air gets thinner the higher you get, the proportions of the gases making it up stay the same, because of the constant mixing. If the air wasn't continually being stirred up, the heavier gases would sink to the lower levels.

OXYGEN REQUIREMENTS

The oxygen to be carried, and the people to whom masks should be made available, varies with altitude, rate of descent and MSA. The latter two are dependent on each other, in that it's no good having a good rate of descent if your safety altitude stops you. It may well be that, although you're flying at a level that requires fewer masks, the MSA may demand that you equip everybody. Preflight stuff includes ensuring that oxygen masks are accessible for the crew, and that passengers are aware of where their own masks are. Check the security of the circular dilution valve filter (a foam disc) on all of them, together with the pressure. Beards will naturally reduce their efficiency. Briefings should include the importance of not smoking and monitoring the flow indicator. All NO SMOKING signs should be on when using it. If you know you will need oxygen at night, start using it from takeoff.

There are three types of oxygen supply, *continuous flow, diluter demand* and *pressure demand*. For a diluter demand system, the *regulator* controls the amount of pure oxygen mixed with air. You start to become physically affected from lack of oxygen above 6,000 ft, so smokers are at a disadvantage before they start.

The Gas Laws

The atmosphere behaves like any other gas, and obeys all the physical laws, such as expanding when heated, etc. Temperature, pressure and humidity all affect density, which ultimately affects aircraft performance.

Charles' Law states that temperature is directly related to volume. *Boyle* discovered that pressure is inversely related to it, and *Dalton* says that the total pressure of a mixture of gases is the same as the sum of the *partial pressures* of the gases it is made of, which is relevant when it comes to dealing with oxygen. In other words, each gas's pressure contributes a part of the total pressure, according to its constituent proportion. Thus, there are three variables when it comes to gases - *pressure, density* and *temperature*, which are all intimately related. For example, if a gas were restrained in a rigid container, increasing the temperature would increase the pressure, and *vice versa*. If the container were not rigid, the density could vary instead.

At sea level, a gas has a third of the volume it would have at 27,500 feet.

Pure oxygen is a colourless, tasteless, odourless and non-combustible gas that takes up about 21% of the air we breathe. Although it doesn't burn of itself, it does support combustion, which is why we need it, because the body turns food into heat. As we can't store oxygen, we survive from breath to breath. How much you use depends on your physical activity and/or mental stress - for example, you need 4 times more for walking than sitting quietly. The proportion of oxygen to air (21%) actually remains constant, but as the air gets less dense, each lungful contains less oxygen in proportion (that is, the partial pressure becomes less), which is why high altitude flight requires extra supplies. Nothing more is required below 5000 feet, as 95% of what you would find on the ground can be expected there. However, at over 8000 feet, you may find measurable changes in blood pressure and respiration, although healthy individuals should perform satisfactorily. Lack of oxygen leads to.......

Hypoxia

A condition where you don't have enough oxygen in the tissues, from inefficient transfer of it into the blood, but anaemia can produce the same effect, as can alcohol (there are several types of hypoxia, but we won't bother with that here). In other words, there may really be too little oxygen, or you don't have enough blood to carry what you need around the body - you may have donated some, or have an ulcer. You might also be a smoker, with your haemoglobin blocked by carbon monoxide (*anaemic hypoxia*). A blockage of 5-8%, typical for a heavy smoker, gives an equivalent altitude of 5000-7000 feet before you even get airborne! The effects of hypoxia are similar to alcohol but the classic signs are:

- *Personality changes.* You get jolly, aggressive and less inhibited

- *Judgement changes.* Your abilities are impaired; you think you are capable of anything and have much less self-criticism

- *Muscle movement.* Becomes sluggish, not in tune with your mind

- *Short-term memory loss.* This leads to reliance on training, or long-term memory

- *Sensory loss.* Blindness occurs (colour first), then touch, orientation, hearing

- *Loss of consciousness.* You get confused first, then semi-conscious, then unconscious

- Blueness

The above are *subjective* signs, in that they need to be recognised by the person actually suffering from hypoxia, who is in the wrong state to recognise anything. External observers may notice some of them, but especially lips and fingertips turning blue and possible hyperventilation (see below) as the victim tries to get more oxygen. However, the normal reaction to lack of oxygen, e.g. panting, does not appear, because there is no excess CO_2. As with carbon monoxide poisoning, the onset of hypoxia is insidious and can be recognised only by being very aware of the symptoms, which are aggravated by:

- *Altitude*. Less oxygen available, less pressure to keep it there
- *Time*. The more exposure, the greater the effect
- *Exercise*. Increases energy usage and hence oxygen requirement
- *Cold*. Increases energy usage and hence oxygen requirement
- *Illness*. Increases energy usage and hence oxygen requirement
- *Fatigue*. Symptoms arise earlier
- *Drugs* or *alcohol*. Reduced tolerance
- *Smoking*. CO binds to blood cells better than oxygen

The *times of useful consciousness* (that is, from the interruption of the oxygen supply to when you can do nothing about it) are actually quite short:

Height	Time
18 000'	30 mins
22 000'	4-8 mins
25 000'	2-3 mins
30 000'	30-60 secs
35 000'	15-35 secs
45 000'	8-12 secs

Hyperventilation

This is simply overbreathing, where too much oxygen causes carbon dioxide to be washed out of the bloodstream, where the plasma gets too alkaline, and the arteries reduce in size, meaning that less blood gets to the brain (oxygen is actually quite corrosive - it belongs to the same chemical family as chlorine and fluorine, so too much is toxic). Unconsciousness slows the breathing down so that the CO_2 balance is restored, but falling asleep is not often practical! The usual cause is worry, fright or sudden shock, but hypoxia can be a factor - in fact, the symptoms are similar to hypoxia and include:

- Dizziness

- Pins and needles, tingling

- Blurred sight

- Hot/Cold feelings

- Anxiety

- Impaired performance

- Loss of consciousness

The last one is actually one of the best cures, since the body's automatic systems take over to restore normality. If in doubt, treat for hypoxia.

Pressure Changes (Barotrauma)

Aside from oxygen, the body contains gases of varying descriptions in many places; some occur naturally, and some are created by the body's normal working processes. The problem is that these gases expand and contract as the aircraft climbs and descends (*Boyle's Law*), and live in various cavities, such as the sinuses. Some need a way out, and some need a way back as well:

- Gas in the ears is normally vented via the Eustachian tubes. If these are blocked (say with a cold), the pressure on either side of the eardrum is not balanced, which could lead (at the very least) to considerable pain, and (at worst) a ruptured eardrum

- Sinus cavities are also vulnerable to imbalances of pressure, and are affected in the same way as eardrums

- Gas in the gut can be vented from both ends

- Teeth may have small pockets of air in them, if filled, together with the gums. Although dentists nowadays are aware of people flying, and pack fillings properly, the general public don't fly every day, as you do, so it's best to be sure. High altitude balloonists actually take their fillings out

Motion Sickness

This is caused by a mismatch between the information sent to the brain by the eyes and ears. Accelerating from straight and level flight may give the impression of pitching up, because the sensors in the inner ear perceive the body weight as going rearwards and downwards. As the most dependable source of sensory information is your eyes, believe your instruments.

Decompression Sickness

Where pressures are low, nitrogen in the blood comes out of solution (typically above 18,000', but more so at 25,000'). Bubbles can form, and are especially painful in the joints (e.g. the bends, for the joints, the creeps (skin), chokes (lungs) and the staggers (brain). All this derives from Henry's Law.

Unfortunately, these bubbles do not redissolve on descent, so if you are affected you may need to go into a decompression chamber. For this reason, diving before flight should be avoided, as extra nitrogen is absorbed while breathing pressurised gas and will dissolve out as you surface again. Don't fly for 12 hours if you have been underwater with compressed air, and 24 hours if you've been below 30 feet.

Loss of Pressurisation

Put on oxygen masks and select 100% oxygen, with the aircraft in a rapid and controlled descent to at least 10,000 ft cabin altitude. Seat belt and no smoking signs should be turned on, and cabin crew should return to their seats. If you cannot get lower, be alert for decompression sickness, as it may occur even with a good enough supply of oxygen (cabin altitudes may be as much as 5000 ft higher than the actual altitude, because of the venturi effect as the air is sucked out of the cabin). Very rarely, lung damage can be caused by rapid depressurisation, which can be avoided by breathing out.

THE CIRCULATORY SYSTEM

This is made up of the heart, arteries, arterioles, capillaries, veins and blood.

The Heart

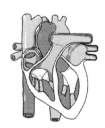

This item consists of four chambers, including *ventricles*, which are actually what contracts when the heart pumps. Blood goes from the *right* ventricle through the *pulmonary artery* to the lungs. Blood on its way back collects in the *left atrium* from where it is injected into the *left* ventricle, which shoots it through to the *dorsal aorta*. Both ventricles and both atria contract together.

The rate of contraction, or the *pulse rate*, is around 72 beats a minute when at rest. As the ventricle pumps about 70 ml of blood per beat, the *cardiac output* is about 5 litres a minute (actually between 4.9-5.3).

Arterial blood pressure is sensed by bundles of nerves in cavities called *sinuses*. There are two in the main arteries to the brain, and another on the aorta, the *carotid* and *aortic sinus pressoreceptors*, respectively, but you knew that already. The brain varies secretions of two hormones in response to their signals, to regulate blood pressure by narrowing the arteries.

The *systolic blood pressure* is the peak pressure as blood is pumped from the heart to the aorta. The *diastolic pressure* is the lowest, produced when resting between beats. Normal blood pressure lies between 110-145 mmHg (systolic) and 70-90 mmHg (diastolic) - you might see a figure of 120/80 (120 over 80). However, standard values are 100 and 60 mg, or 100/60, with the limits regarded as 160 and 100 mmHg, or 160/100. When older, the systolic pressure should be roughly 100 plus your age in years. The arterial pressure in the upper arm is equivalent to the pressure in the heart, which is why it is used in medical examinations. The heart does not rest in the same way as other muscles - instead, it take a mini-rest for a microsecond or two in between beats.

Blood

55% colourless plasma, for transporting CO_2, nutrients and hormones, and 45% cells. *Red cells* transport oxygen via haemoglobin, and *white cells* (*leukocytes*) fight infection. *Platelets* are for clotting blood. CO_2 in solution forms a weak carbonic acid which also helps to maintain the blood's acid balance. The amount of haemoglobin in the blood depends on the amount of oxygen in the *lungs*. Reductions in the amount of haemoglobin available reduces the blood's ability to transport oxygen (to cause anaemia). This could arise from either less red blood cells or the concentration of haemoglobin in them. *Anaemia* means that there are too few red blood cells, and a limited capacity to transport oxygen (more iron often cures it). *Anaemic Hypoxia* is the lack of oxygen resulting from anaemia.

Heart Disease

Heart disease can be grouped into 3 categories:

- *Hypertensive* - from high blood pressure, working the heart harder so it gets enlarged (anxiety, etc.).

- *Coronary*, or *Arteriosclerotic* - hardening of arteries through excessive calcium, or cholesterol, which again makes the heart work harder (bad diet).

- *Valvular* or *rheumatic* - where valves are unable to open or close properly, allowing back pressure to build up (old age).

A heart attack would lead to *circulatory shock*, or a failure of the blood supply. Adrenaline increases the speed and force of the heart beat. To reduce the risks of heart disease, double your resting pulse for at least 20 minutes 3 times a week (a recent US study has suggested that this will also lengthen your life by around two years - which is spent running!)

GENERAL HEALTH

The body's main fuel is glucose, which can either be converted from different types of food, or eaten directly. Levels of glucose are regulated by the *pancreas*, which secretes *insulin* to reduce blood sugar levels by getting it into cells or converting it into fat if there's no room.

However, sugar is one of the most harmful substances we can put into our bodies on a daily basis, and there is almost no processed food that does not contain it - even baked beans. Certainly, there is hardly a cereal product without it (did you ever wonder why cereals are fortified with vitamins? It's because they have them all taken out first!) Sugar that is not needed to maintain adequate glucose levels and replenish stored glycogen in the liver and muscles is converted to fat, helped along by insulin, which also tends to block the conversion of fat back to glucose, so a high insulin level makes it difficult to remove the fat it created in the first place. The problem is that, on the average Western diet, our insulin levels are almost permanently high, which is something that our bodies are simply not built to cope with - the pancreas needs a rest! Thus, we should try to eat so that large spikes of insulin are not generated, which can be difficult in a normal pilot's lifestyle. That is, insulin should be injected into the bloodstream under more controlled conditions - processed foods are converted into glucose very quickly, which is the real problem. The type of carbohydrate you eat will determine how this happens. After reading most of the diet books around, I have come to the following conclusions:

- It is not necessarily the fat you eat, but the fat created from sugar that is bad for your health

- Don't eat anything processed - which is usually anything "white", or at least with white flour in

- Eat fruit by itself - although fruit contains sugars, they also contain enzymes and other beneficial substances, and don't stimulate so much insulin (around a third). However, once you combine fruit with other food, you get the full non-benefit. Also, fruit is digested mostly in the small intestine, and eating it after a large meal causes this to be delayed, with fermentation that causes indigestion

- If you drink alcohol (in moderation!), dry (low sugar) red wine is best

- Exercise, but not so much that you need to eat a lot to produce the glucose you need

- Don't eat a heavy meal just before going to bed

- Drink lots of fluid (not with sugar or caffeine in! Caffeine can pull calcium out of your bones)

- Eat lots of fibre

- Eat more water-based food, such as fruit, greens, tomatos, etc. in their raw state - try for around 70% of your total diet

Hypoglycaemia

The most common problem (in the normal pilot's lifestyle, anyway) is low blood sugar, caused by missed meals and the like. Although you may think it's better to have the wrong food than no food, be careful when it comes to eating choccy bars in lieu of lunch, which will cause your blood sugar levels to rise so rapidly that too much insulin is released to compensate, which drives your blood sugar levels to a *lower* state than they were before - known in the trade as *rebound hypoglycaemia*. Apart from eating "real food", you will minimise the risks of this if you eat small snacks frequently instead of heavy meals after long periods with nothing to eat. Complex carbohydrates are best, in the shape of pasta, etc.

Hypoglycaemia is fair enough in the short term, but long-term can be a disease. Although not life threatening, it is a forerunner of many worse things and should be looked at. The important thing to watch appears to be the suddenness of any fall in blood sugar, and a big one can often trigger a heart attack. A high protein diet will tend to even things out, as protein helps the absorption of fat, which is inhibited if too much insulin is about. Warning signs include shakiness, sweatiness, irritability or anxiety, difficulty in speaking, headache, weakness, numbness or tingling around the lips, inability to think straight, palpitations and hunger.

At its worst, hypoglycaemia could result in coma, but you could also get seizure and fainting. Eat more if you exercise more.

Hyperglycaemia

This is the opposite of the above, an *excess* of blood sugar. Symptoms include tiredness, increased appetite and thirst, frequent urination, dry skin, flu-like aches, headaches, blurred vision and nausea. This condition causes dehydration, so have fluids around. Also, decrease stress.

Diabetes

Glucose in the blood provides energy, but sometimes the body cannot either produce the insulin in the first place that is required to get it into the cells (Type 1 diabetes), or make use of it in the best way (Type 2). The former tends to appear in people under 40, and the latter in people over 40, although they can occur in either. Both can be treated by diet and/or insulin injections. Exercise is good, too, as it uses up blood sugar that would otherwise need insulin to get rid of it. Since insulin is needed to get glucose into your cells, it follows that, if this is not done, blood sugar levels can be dangerously high. In the short term, you may be tired, thirsty, urinate frequently (and get dehydrated), have blurred vision, headaches and tingling fingers and toes. More long-term, blood vessels can get damaged from high blood pressure and cholesterol levels (so important is this, that, if you were on an island with a bottle of blood pressure pills and a bottle of diabetes pills, you'd better take the blood pressure pills first! High blood pressure will push proteins through kidney walls, to be found in urine).

Alcohol

Whilst nobody should object to you taking a drink or two the evening before a flight, you should remember that it can take over 3 days for alcohol to clear the system (it remains in the inner ear for longest). Within 24 hours before a planned departure, you should not drink alcohol at all; certainly not on standby. The maximum blood level is officially .2 mg per ml, a quarter of the driving limit in UK, but it's not only the alcohol that causes problems - the after-effects do as well, like the hangover, fatigue, dehydration, loss of blood sugar and toxins caused by metabolisation. The body produces its own alcohol - around .04 mg per 100 ml.

Although it appears otherwise, alcohol is not a stimulant, but an anaesthetic, which puts to sleep those parts of the brain that deal with inhibitions - the problem is that these areas also cover judgement, comprehension and attention to detail. In fact, the effects of alcohol are the same as hypoxia, dealt with elsewhere, in that it prevents brain cells from using available oxygen. One significant effect of hypoxia in this context is the resulting inability to tell that something is wrong. It takes the liver about 1 hour to eliminate 1 unit of alcohol from the blood (officially, alcohol leaves the body at 15 milligrams per 100 ml of blood per hour). 1 unit is considered to be the same as 1 measure of spirit, a glass of wine or half a pint of beer, although this has recently changed. Based on old units, however, the number of units per week beyond which physical damage is likely is 21 for men and 14 for women.

The reason why the world spins round after a few drinks is because alcohol diffuses into the fluid in the semicircular canals, which reduces its density and makes it more prone to slosh around and keep on moving.

As far as passengers are concerned, although they get cabin service, persons under the affluence of incohol or drugs, of unsound mind or having the potential to cause trouble should not be allowed on board - certainly, no person should be drunk on any aircraft (people aren't generally aware that one drink at 6000 feet is the same as two at sea level). This is not being a spoilsport - drunks don't react properly in emergencies and could actually be dangerous to other people (which is why I always get an aisle seat - I don't have to get round people in the way). So, it's not just for their own good, but that of others as well. If you need to get rid of obstreperous passengers, you can always quote the regulations at them (or even use sarcasm), but don't forget to fill in an Occurrence Report.

The UK ANO says that flight crew members may not be under the influence of drink or a drug to such an extent as to impair their ability to act as such, which presumably means that if they are alcoholic and fly better with a snort or two under their belt, they may drink and fly (joke!). JAR OPS, on the other hand (1.085 and 3.085), states that crew members must not perform duties:

- Under the influence of any drug that may affect faculties contrary to safety

- Until a reasonable time has elapsed after deep water diving

- After blood donation, except after a reasonable time

- If in any doubt of being able to accomplish assigned duties

- If knowing or suspecting that they are suffering from fatigue, or feel unfit so that the flight may be endangered

Crew members must not:

- Consume alcohol less than 8 hours before the specified reporting time for flight duty or the start of standby

- Start a duty period with blood alcohol levels over 0.2 promille*

- Consume alcohol during the flight duty period or while on standby

*Less than 0.2 parts per thousand, around 20 mg of alcohol per 100 ml of blood.

Australian researchers compared the effects of alcohol and fatigue on performance. They found that being awake for 17 hours in a row is as bad as a blood alcohol concentration of .05%. The legal limit for driving in Canada, for example, is .08%.

Note: Most regulations are worded so that you can have a drink 8 hours before reporting time, but you must also not have alcohol in your system.

Exercise

Physical exercise strengthens the heart and enhances the blood flow, which helps when you get hypoxia. More of this below.

Medications

Although the symptoms of colds and sore throats, etc. are bad enough on the ground, they may actually become dangerous in flight by either distracting or harming you by getting more serious with height (such as bursting your eardrums, or worse). If you're under treatment for anything, including surgery, not only should you not fly, but you should also check that there will be no adverse effects on your physical or mental ability, as many preparations combine chemicals, and the mixture could make quite a cocktail. No drugs or alcohol should be taken within a few hours of each other, as even fairly widely accepted stuff such as aspirin can have unpredictable effects, especially in relation to hypoxia (it's as well to keep away from the office, too - nobody else will want what you've got). Particular ones to avoid are antibiotics (penicillin, tetracyclines), tranquilisers, antidepressants, sedatives, stimulants (caffeine, amphetamines), anti-histamines and anything for relieving high blood pressure, and, of course, anything not actually prescribed. Naturally, you've got to be certifiable if you fly having used marijuana, or worse.

ANAESTHETICS

All procedures requiring local or regional anaesthetics disqualify you for flying for at least 12 hours.

DRUGS

Although altering your state of consciousness is one way of dealing with stress, using drugs is not the way to do it. As with all short cuts, there are unwanted side effects. A study of airline pilots landing in a simulator found that performance was significantly impaired up to 24 hours after smoking one marijuana cigarette with 19 milligrams of THC, although the pilots thought they were doing OK. Marijuana also affects short-term memory and the the transfer of information into long-term memory.

Diseases

You often fly to places with very low standards of hygiene and/or disease-carrying insects. Although not good for you in the long run, on such occasions, processed and packaged food can be a real lifesaver, as can bottled water. As with many things in life, prevention is better than cure - for malaria, certainly, a good tactic is to avoid being bitten (by the female Anopheles mosquito) in the first place, by wearing appropriate clothing, even though it is hot. Yellow fever is also insect-borne, but can be vaccinated against. You get cholera from dirty water.

Blood Donations

Pilots generally are discouraged from giving blood (or plasma) when actively flying, because a donation may lead to a reduced tolerance of altitude. Some dental anaesthetics can cause problems for up to 24 hours or more, as can anything to do with immunisation. If you do give blood, try to leave a gap of 24 hours, including bone marrow donations. Although blood volume is restored in a very short time, and for most donors there are no noticeable after-effects, there is still a slight risk of faintness or loss of consciousness. After a general anaesthetic, see the doctor.

Food poisoning can also be a problem, and not just for passengers - the standard precaution (like in *Airplane!*) is to select different items from the rest of the crew, even in the hotel, or at least eat at different times.

Don't forget to inform the authorities (in writing) of illnesses, personal injuries or presumed pregnancies that incapacitate you for more than 21 days (in UK). Pilots involved in accidents should be medically examined before flying again.

INCAPACITATION

In the obvious case, you could collapse and fall across the controls. Less noticeable is the sort of incapacitation that comes with boredom or lack of mental stimulation on longer trips, where you may physically be in the cockpit but mentally miles away. Even disorientation during instrument flight is included. Incapacitation can be *gradual* or *sudden*, *subtle* or *overt*, *partial* or *complete* and you may not have any warning.

Partial or Gradual

This bit concerns medical symptoms that affect your handling ability, to the extent that you would have to hand over control when multi-crew, but would otherwise have to land quickly when by yourself. These might include severe pain (especially sudden severe headache or chest pain), dizziness, blurring or partial loss of vision, disorientation, vomiting or diarrhoea (airline food again!) Temporary symptoms often indicate something more severe, so don't be tempted to get back in the air. You should react before any illness becomes severe enough to affect your handling, so an immediate radio call is essential. The first consideration must be for the safety of the passengers, so medical assistance for you must be a lesser priority, though the former may well depend on the latter.

Sudden or Complete

This may be subtle or overt, and give no warning; Murphy's Law dictates that fatal collapses occur during approach and landing, close to the ground. Detection of subtle incapacitation may be indirect, that is, only as a result of some expected action not being taken, so when you die maintaining your body position, another pilot may not even notice until the expected order of events becomes interrupted.

G TOLERANCE (ACCELERATION)

The body can only cope with certain amounts of G-force, from the effects of *acceleration*, which increase your weight artificially. When there is none, you are subject to 1G. *Linear acceleration* is what you get in crashes, *radial acceleration* while turning, and *angular acceleration* when the rate of rotation changes, and which affects your sense of balance.

Negative G acts upwards and can increase the blood flow to the head, leading to *red out*, facial pain and slowing down of the heart. In addition, your lower eyelids close at -3G. Positive G is more normal, but will drain the blood, with the obvious consequences, including loss of vision, called *grey out*, at +3.5 G This could end up as *black out* and unconsciousness at +6 G. Both are affected by hyperventilation, hypoxia, heat, hypoglycaemia, smoking and alcohol, all discussed more fully below. The *valsalva manoeuvre* can be used to help cope with high G (close your throat and hold your nose), or you could use a pressure suit.

The body can tolerate 25G vertically and 45G horizontally - if you don't wear shoulder straps, tolerance to forward deceleration reduces to below 25G, and you will jackknife over your lapstrap with your head hitting whatever is in front of it at 12 times the speed it is coming the other way.

NOTES

PSYCHOLOGICAL FACTORS

O r, in other words, that which influences, or tends to influence, the mind or emotions. We are concerned with them, because they can influence the way we interpret information on which we base decisions. With optical and aural stimuli (discussed in Chapter 2), *the processing is done in the brain*, which uses past experience to interpret what it senses - it therefore has expectations, and can pre-judge a situation. In fact, as accident reports routinely show, in high stress conditions, the brain may blank out information not directly concerned with the task in hand.

PERCEPTION

This is the process of giving meaning to what is sensed. *Set* is a tendency to respond in a certain way, in line with expectations based on past experience - *perceptual set* relates this to the perception process. Variations on this theme could come from:

- The stimulus itself. For example, the moon at the horizon appears larger than the moon when overhead, even though the image on the retina will be the same, because many of the visual cues for greater distance occur when it is viewed near land.

- The situation, or the context in which an image is viewed. The figures 1and 3 could be seen as the letter B if they were included together in a list of letters.

- The state of the perceiver with regard to motivation or emotion. If you are hungry, pictures of food can appear to be brighter.

LEARNING

In simple terms, this can be defined as a long-term change in behaviour based on experience, whether its other peoples' (reading, studying) or your own. *Reversion* can occur after a pattern of behaviour has been established, mainly because you get so used to doing it - you may accidentally carry out a procedure you have used for years, even after revision, which is quite possible under stress, but it can happen under normal circumstaces - a

Beaver pilot went back to fly the piston-engined variety after a long spell on the turbo version, and ran the tanks dry before changing over, which is a normal practice in the bush, but it caused quite a stir among the passengers in the airline he was flying for. There are several behaviour patterns (or pilot performance levels) concerning this:

- *Skill-based learning* is based on practice and prior learning, to become part of the "muscle memory", or motor programs, of your body (as when learning the piano), meaning that reactions are largely unconscious and automatic. As such it does not require conscious monitoring, but it can lead to *environmental capture*, that is doing something because it's always done and not because it's the right thing to do (saying "3 greens", for example, without lowering the gear). You could also end up with the right skill in the wrong situation (*action slip*), meaning pulling the flap lever instead of the gear, but you might also not catch new stimuli when operating largely in automatic mode.

- *Rule-based learning* is that which relies on previously considered courses of action, and follows procedures, like checklists and SOPs, so demands a little more conscious monitoring if the rules are to be followed in the right sequence. It is kept in long term memory, requiring the *decision channel* and working memory for execution - an inexperienced pilot may have a problem with this if the rules are imprecise and assume a minimum level of knowledge for them to be used properly. What usually happens when an accident occurs is that the brain goes smartly into neutral whilst everything around you goes pear-shaped. Checklists can help to bridge the gap of inactivity by giving you something more or less correct to do whilst psyching yourself up and evaluating information ready for a decision. The US Navy, for example, trains pilots to stop in emergencies, and reset the clock on the instrument panel, which forces them to relax, or at least not to panic. Rule-based behaviour is generally robust, which is why procedures and rules are important in maintenance. However, you can use the wrong procedure as a result of misdiagnosis (of course, you could always forget the procedure).

- *Knowledge-based learning* relies on previous experience. It's the sort of stuff you apply if you need to think things through, or maybe work on the *why* so the *how* becomes apparent. Inexperienced pilots are more likely to make knowledge-based mistakes when they are forced into such behaviour in situations that have not been encountered before. Thus, errors at this level arise from making diagnoses without full knowledge of a system. When problem-solving, the transition from rule-based to knowledge-based activities is governed by the unsuitability of known rules for the problem posed.

Errors can occur at all three levels (see *Errors*, elsewhere).

Factors That Affect Learning

There are, in theory, three stages, starting with a theoretical knowledge of what needs to be done, through practice, to where the knowledge is completely in memory (although I have never felt that learning the complete alphabet was necessary before learning to read). In aviation terms, however, in the first (or *cognitive*) phase, an instructor might talk about skills which you are to acquire, including the task, typical errors and target performance. Next comes the *associative phase*, where techniques are demonstrated and learned, and errors are gradually reduced. The *Autonomous* or *Automatic* stage is where you have it down pat. These factors may help in getting an instructor's message (for example) across:

PERSONALITY

Being polite and respectful. Students are intelligent - they just don't know what they are about to be told (kids are way smarter than most adults, too, but they can't speak English, and people often think they're stupid).

THE ENVIRONMENT

Think back to school. Remember all those distractions that stopped you listening to the teacher? Enough said!

READY, STEADY, GO! (MOTIVATION)

People need to be ready to learn, and voluntary students already have motivation, but it won't help if you haven't prepared properly, you rush them, or they have other problems that belong outside the classroom. Anxiety will affect this.

BUT WHY?

It's a great help to explain *why* some subjects are being taught, as well as their relationship in the great scheme of things. As an example, at school I had trouble with simultaneous equations, because I couldn't see the point of them. It wasn't until a new teacher arrived, an ex-bomber pilot, who explained that, if you had two aeroplanes, and knew the wingspan of only one of them, you could find the wingspan of the other with a simultaneous equation (you can also use them to calculate hyperbolic differences in Area Navigation, but that's not important right now). All of a sudden, it was lot simpler. If people know how or why things work, even if they don't know the answer immediately, they can usually work it out.

REVIEW (RECENCY)

It has been found that, within two days, if it isn't reviewed, people remember less than 70% of any subject matter they may have studied. By the end of the month, the figure falls to 40%. On the other hand, if it's looked over again within 2 days, then 7, you should be above the 70% level until the 28th day. Another review then should make it remain long-term. In fact, short and frequent bursts of study are more effective than one long one - the brain appears to like short "rests" to assimilate knowledge. Constant reviewing is the key, especially for a short time at the end of each day. (Source: *Ohio State University*).

RIGHT FIRST TIME

Do not teach something, and change it later! Whatever you teach *must* be done right first time, because that is the impression that sticks in the student's mind - if you've ever taught anyone to use a computer, you will know exactly what I mean. When learning things for the first time, people expect the situation to be the same *every time*. If you demonstrate something, then say "No, that's wrong" then have to sort it out in front of them, they will not only get confused, but you will also look pretty stupid. Lessons and demonstrations must be learnt until they are perfect, then the students have to be monitored to make sure they do it right as well, otherwise they will just learn bad habits, and their future Training Captains will be after you!

LOGICAL PROGRESSION

The components of a lesson should be in a logical sequence, preferably with a link between one section or another. If you listen carefully to radio broadcasters, you will realise that they try to segue each subject they talk about as seamlessly as possible - in fact, it is a matter of professional pride to take widely differing subjects and still produce a link between them, however unlikely. It isn't a hard skill to learn.

PRACTICAL

Mental processes are reinforced by physical means. It used to be the custom to take the young lads of an English village around its boundaries and beat them with a switch to make sure they knew where they were (this was called "beating the bounds"). I'm not suggesting that you should beat your students (well......), but it does make the point. If you can, reinforce the stuff you are blethering about with a physical exercise.

INTENSITY OF EXPERIENCE

Just because you are teaching aviation matters, don't be shy of using examples from other walks of life to reinforce a point. For instance, when teaching people to use a wordprocessor, I often use my aviation experience as part of the lesson - the most dramatic is to have them type a few pages of text, then pull the plug on the power supply, ensuring that they lose all their work. This "practice engine failure" is most instrumental in emphasising the importance of saving their work regularly. You could do the same in a non-threatening aviation situation. You need to find one that is drastic enough to gain their attention, but that will not lead to danger.

THE FEEL GOOD FACTOR

This involves making people feel good about themselves as an aid to learning, mainly by not concentrating on the negative, but the positive, although this doesn't mean that you shouldn't point out their mistakes. Children with learning difficulties, for instance, are often taken swimming, something they can do well, to reinforce their self-image and encourage them with the more challenging aspects of their lives. If one of your students is having a problem, change the subject for a short while to something they excel in, then come back to the previous one. Always point out what they did well before mentioning what they did wrong, and even then introduce it in as positive a manner as possible. Why not get the student to contribute to the lesson somehow? Instead of just talking, listen to them - I guarantee you will learn a lot. Their feedback is important to your own development as an instructor.

TECHNIQUE

Poor instructional ability and communications skills.

AGE

Learning may not be as easy after the age of about 60.

STRESS
• •

Flying is stressful, there's no doubt about that, but should stress be a problem? It's arguable that a little is good for you; it stops you slowing down and keeps you on your toes; this is the sort associated with success. Excessive stress (or *dis*tress), on the other hand, in the form of pressure (that is, stress without respite) can lead to fatigue, anxiety and inability to cope, and is associated with frustration or failure. Long term stress can affect your immune system, and, over the course of a career, lead to failing medicals earlier than you should. There is a large body of evidence to indicate that stress causes many modern illnesses, certainly headaches, asthma, hypertension and heart disease, and a reduction in efficiency of the immune system. In short, it disturbs the body's *homeostasis*, which is like the body's thermostat when it comes to its comfort and efficiency.

You can measure stress by looking for the amount of cortisol in the urine. Cortisol is produced by the adrenal gland, which can become enlarged under prolonged stress.

Stress and preoccupation have their effects; a PA31 pilot was doing a cargo flight with three scheduled stops, but he did not refuel or even shut down at any of them, so both engines stopped after the last delivery. He was anxious to get home as his wife was in hospital. This illustrates some of the four sources of stress, which are:

- Personality

- Family

- Occupation

- Situation

and their combinations.

Fight or flight responses are bodily changes that prepare it for action - adrenaline starts to pump and many other changes take place as well, including a rise of sugar and fats in the blood (including cholesterol, from the liver), endorphins (from the hypothalamus), faster respiration, thicker blood (to carry more oxygen), tense muscles and the stopping of digestion, so more blood can be diverted to where it is required. All this happens very quickly, but it cannot be maintained for long - if it is, the body can be adversely affected.

When under stress you may well revert to former training - watch out for those levers in the wrong place on the new machine (e.g. *reversion*)! An overstressed pilot may show mental blocks, confusion, channelised attention, resignation, frustration, rage, deterioration in motor coordination, and fast speaking in a high-pitched voice. This is because, with stress, the blood flows from the brain to the muscles, which affects the decision-making process.

The reason why stress is a problem for the human body is evolutionary - the autonomic nervous system, if you remember, works as a whole unit, which may have been OK when coming across a mammoth, but not for the more everyday stuff we have to cope with today.

What is Excessive Stress?

Anything that has a sufficiently strong influence to take your mind off the job in hand, or to make you concentrate less well on it. Not only are you not doing your job properly, but subconsciously feel guilty about it, too, which is enough to set up a little stress all of its own. We all like to feel we are doing the best we can possibly do, and it disturbs our self-image to feel that we're not. Consequently we get angry at ourselves for being in such a position, which increases the stress, which further takes us away from the job, and so on.

Common situations causing stress include grief, divorce, financial worries, working conditions, management pressure, pride, anger, get-home-it is, motivation (or lack of), doubts (about abilities, etc), timetable, passengers' expectations, etc. There are many types of difficult people, but the most common you will encounter is the bully, from both ends of the spectrum (it has been known for Captains to get violent). Bullies are insecure, and jealous. Anyhow, under stress, because you cannot concentrate, your judgement becomes impaired, and you make rash decisions, just so the problem can go away. Mainly, though, you lose perspective.

All the above leads to anxiety, which is really based on fear, if you think about it (fear of people not liking you, of losing your job, etc.). As anxiety can cause stress, you get a circulating problem. People have their own ways of dealing with stress, so what works for one does not necessarily work for someone else. This is possibly because of the evaluation of the stress that that particular person has, i.e. whether they feel they can cope and their perception of the problem. It is *perception* of demands and abilities, rather than actual problems that affect the individual. If you *feel* you are capable, or in control, your stress level will be relatively low.

Symptoms of stress include:

- Anxiety and apprehension, depression, gloom, mood swings
- Detachment from the situation
- Failure to perceive time
- Fixation of attention
- Personality changes
- Voice pitch changes
- Desire for isolation
- Reduced cognitive ability
- Poor emotional self-control
- Unsafe cavalier attitude
- Anger

The most distinguishing feature is the amount of control you may or may not have. The more helpless you feel, the more stress you have.

FATIGUE

Prolonged exposure to fatigue can reduce the capabilities of your immune system and make you quite ill. Like the frog in a saucepan of warm water that is getting hotter, you don't usually notice until you collapse in a heap at the end of the flying season. If you're like me, you then need about 6 months to recover. Helicopter pilots are especially prone to fatigue, due to the high workload and intense decision making, and vibration (see *Whole-body vibration*, below). In fact, in Canada, the government compares four hours' worth to eight hours' hard labour and double that when longlining (not including the normal A-B stuff, of course, and naturally, fixed wing pilots engaged on short sector work have similar strains imposed, without the vibration).

Fatigue is typically caused by delayed sleep, sleep loss, desynchronisation of normal circadian rhythms and concentrated periods of physical or mental stress or exertion. Working long hours, during normal sleep hours or on rotating shifts, all produce fatigue to some extent. As mentioned elsewhere, Australian researchers compared the effects of alcohol and fatigue on performance. They found that being awake for 17 hours in a row is as bad as a blood alcohol concentration of .05%. The legal limit for driving in Canada, for example, is .08%.

Symptoms of fatigue may include:

- diminished perception (vision, hearing) and a general lack of awareness

- diminished motor skills and slow reactions

- problems with short-term memory

- channelled concentration - fixation on single possibly unimportant issues, to the neglect of others and failing to maintain an overview

- being easily distracted by unimportant matters

- poor judgement and decision-making, leading to increased mistakes

- abnormal moods - erratic changes, depressed, periodically elated and energetic

- diminished standards

Most people need about 8 hours' sleep, and you can do with less for a few days, creating a *sleep deficit*, but it's not only the amount of sleep you get, but *when* you get it that counts, so fatigue is just as likely to result from badly planned sequences of work and rest, or being too long away from base without a day off. A surprising amount (over 300) of bodily functions depend on the cycle of day and night - we have an internal (*circadian*) rhythm, which is modified by such things, which, oddly enough, is 25 hours. You naturally feel best when they're all in concert, but the slippery slope starts when they get out of line. The best known desynchronisation is jet lag, but it also happens when you try to work nights and sleep during the day. Bright light can fool your body into thinking it's day when it's not. One day for each time zone crossed is required before sleep and waking cycles get in tune with the new location, and total internal synchronisation takes longer (kidneys may need up to 25 days). Even the type of time zone change can matter - 6 hours westward requires (for most people) about four days to adjust - try 7 for going the other way! This Eastward flying compresses the body's rhythm and does more damage than the expanded days going west; North-South travel appears to do no harm.

Symptoms of jet lag are tiredness, faulty judgement, decreased motivation and recent memory loss. They're aggravated by alcohol, smoking, high-altitude flight, overeating and depression, as found in a normal pilot's lifestyle. In view of all this, you have a maximum working day laid down by law, intended to ensure you are rested enough to fly properly.

The two types of fatigue are *acute* and *chronic*, the former being short-term, or more intense, and the latter arising from more long-term effects, like many episodes of acute fatigue, typically found after a long spell of fire suppression. Acute fatigue usually affects the body, and just needs a good nights' sleep to sort things out, whereas the chronic variety can have a mental element, where you might not want to see a helicopter ever again. It typically happens after you've had no rest, food or recreation for some time. Symptoms are insomnia, loss of appetite, and even irrational behaviour. To control its effects, try rest, exercise and proper nutrition.

Foods low in carbohydrate or high in protein help fight fatigue, especially "healthy" ones, like fruit or yoghurt, or cereals, such as granola. Coffee, of course, contains caffeine, which keeps you awake (as does tea), but too much can lead to headaches and upset stomachs. People who drink unleaded coffee (decaffeinated) still report unpleasant side effects, as the process that removes caffeine is allegedly just as harmful, but in different ways. Caffeine has a half-life of about 3 hours, and although it might not stop you getting to sleep, it will affect its quality.

Sleep is actually a state of altered consciousness, in which, although paralysed, you don't lose awareness of the external world, as any mother will tell you (it's actually where your brain focusses internally, as explained in Chapter 5). It is part of a daily cycle which is actually 25 hours long - that is, the sleeping and waking rhythm is about an hour longer than the normal day of 24 hours, which itself is a mean figure anyway (*Moore-Ede, Sulzman & Fuller*, 1982). This is why flying West is easier on the system than flying East - the body's rhythm is extended in the right direction. Various factors, such as cycles of night or day, keep the natural 25-hour tendency in check. Normally, this circadian rhythm works with body temperature, so the body is coolest when it is hardest to stay awake, around 0500. It follows that sleep is harder as body temperature rises.

Types of Sleep

Sleep is a natural state of reduced consciousness involving changes in body and brain physiology, which is necessary to restore and replenish the body and brain. It can be resisted for a short time, but various parts of the brain ensure that, sooner or later, sleep occurs. When it does, it is characterised by five stages that take place over a typical cycle of about 90 minutes, where you descend through 4 stages of NREM (non-REM) sleep, each deeper than the other, and return, when you enter the REM state, where most dreams occur. In *Rapid Eye Movement* sleep, the body and brain become active, heart and metabolism increase and the eyes shift, hence the name. It lasts between 10-25 minutes. All this is an *ultradian* rhythm, which comes from the Latin *ultra dies*, or "outside a day". REM sleep (*Rapid Eye Movement*) refreshes the mind, and *Slow Wave* (NREM) sleep refreshes the body. You are more refreshed if you wake up during the former. Stages 3 and 4 are known as *slow wave* sleep, after the patterns on an EEG, and you are more groggy if you wake up in that period. With REM sleep, the brain is awake in a virtually paralysed body, so you are nearly awake anyway. Most deep sleep occurs earlier in the night and REM sleep becomes greater as it goes on - you might go through 4 or 5 episodes of REM sleep a night. Sleep deprivation experiments suggest that if you are deprived of stage 1-4 or REM sleep, you will show rebound effects, meaning that, in subsequent sleep, you will make up the deficit in that particular type. The nature of fatigue may determine the stages required. For example, you might need extra slow-wave sleep after physical activity.

1 hour of quality sleep equals 2 hours of activity, and you can accumulate up to 8 hours on a credit basis - that is, each sleeping hour gains 2 credit points and each hour awake loses one. Thus, 8 hours' sleep overnight means you will be ready for sleep again 16 hours after waking. If your work pattern is disrupted, you can increase your "credit rating" with a short nap in the back of the aircraft. Alcohol interferes with sleep because of its diuretic action - repeated use disturbs sleep on a long term basis, to give you insomnia. It's worth noting that a shave is about equal to 20 minutes' sleep, in terms of refreshing you, and washing your face or brushing your teeth are also good, as is moving around for 5-10 minutes.

Clinical Insomnia is being unable to sleep under normal conditions. *Situational insomnia* arises out of the circumstances, like sleeping in a strange bed or time zone (*circadian desynchronisation*). Although insomniacs may think they don't sleep at all, they actually spend their time in stages 1 and 2. *Sleep Apnea* stops people breathing for short periods up to a minute, and *Narcolepsy* makes them drop off at any time of the day.

Whole-body Vibration

WBV can be experienced in two ways; through an instantaneous shock with a high peak level (enough to jar you out of your seat, as with turbulence) or through repeated exposure to low levels of vibration from regular motion, as with a helicopter with one rotor blade out of alignment. Its relevance here is its contribution to fatigue (low-frequency vibrations of moderate intensity can induce sleep), but the most common effect of WBV is lower back pain and inflammation that can lead to degeneration of discs or trapped nerves. However, of particular importance to longliners, is the increase in its effects from twisted sitting postures, which increases stress and load on the neck, shoulder and lower back. For example, vibrations between 2.5 and 5 Hz generate strong resonance in the vertebrae of the neck and lumbar region. Helicopter vibration has a peak power at frequencies around 5 Hz.

Vibration from engines can cause micro fractures in vertebrae, disc protrusion, nerve damage and acute lower back pain. Otherwise, short-term exposure has only small effects, such as slight hyperventilation and increased heart rate, plus increased muscle tension from voluntary and involuntary contraction (the tenseness dampens the vibration).

Acute effects (short-term exposures) include:

- Headache

- Chest pain

- Abdominal pain

- Nausea

- Loss of balance

Chronic effects (long-term exposures) include:

- Degenerative spinal changes

- Lumbar scoliosis

- Disc disease

- Degenerative disorders of the spine

- Herniated discs

COPING WITH STRESS

The first thing you can try and do is to eliminate the factors causing your stress. You could also use drugs that act on the autonomic nervous system, but these almost always have side effects and can be addictive, so you also have to deal with withdrawal. I like humour myself, and some people favour eating, exercise or biofeedback machines that help them reduce their heart rate, etc., but most either adjust to the situation, or change it (or walk away from it). However, the willingness to recognise stress and to do something about it must be there; for example, if you don't admit there's a problem at home, there's not much you can do! It is not weakness to admit you have a problem - rather, it shows lack of judgement otherwise. As previously mentioned, it's your *attitude* towards stress that counts, not the situation, as other people may be able to cope with it very well. If you have the usual fight-or-flight symptoms over a relatively minor incident, you are stressed! This energy has nowhere to go and you end up in overdrive, with a very easily ignited short fuse to push you over the edge (see *Anger*).

- *Action Coping* means taking positive action to cope with the source, including removing yourself from the situation, addressing the problem or altering the situation enough to reduce the demands.

- *Cognitive Coping* involves reducing the *Perceived Demand*, maybe by rationalisation or consulting with a friend or colleague. Denial of the problem comes under this heading, but is not recommended.

- *Symptom-Directed Coping* involves treating the symptoms rather than the cause of stress, say by drinking (exercise is better).

FACTORS AFFECTING JUDGEMENT

In short, judgement is process of choosing between alternatives for the safest outcome. It used to be regarded as a factor in the definition of airmanship. Factors that influence the exercise of good judgement include:

- *Lack of vigilance* - keeping an eye on what's going on, is the basis of situational awareness. You need to keep a constant watch on all that is going on around you, however tempting it may be to switch off for a while on a long navex. Monitor the fuel gauges, check for traffic and engine-off landing sites, all the time

- *Distraction* - this is anything that stops you noticing a problem, for example, slowly backing into trees while releasing a cargo net. Keep pulling back from the situation to reaffirm your awareness of the big picture

- *Peer Pressure* - we all like to be liked, whether by people in or outside your own company. Do they want you to fly overweight? Or fly in darkness, even though they are late back? Being too keen to please is part of a self-esteem problem, another aspect of allowing yourself to be put upon

- *Insufficient Knowledge* - although you can look the regulations up in a book, this is not always the most convenient solution, so you need a working knowledge of what they contain, including checklists and limitations from the flight manual, etc. We don't all have the luxury of an aircraft library, or have the time to refer to it if there was one

- *Unawareness of Consequences* - this is an aspect of insufficient knowledge, above. What are the consequences of what you propose to do? Have you thought things out thoroughly?

- *Forgetfulness of Consequences* - similar to the above

- *Ignoring the Consequences* - again, similar to the above, but more of a deliberate act, since you are aware of the consequences of your proposed actions, but choose to ignore them

- *Overconfidence* - this breeds carelessness, and a reluctance to pay attention to detail or be vigilant. Also, it inclines you to be hasty, and not consider all the options available to you. This is where a little self-knowledge and humility is a great help

Fascination

This is where pilots fail to respond adequately to a clearly defined stimulus despite all the necessary cues being present and the proper response available. A study in the 1950s (*Clark, Nicholson, and Graybiel*, 1953) classified experiences in 2 categories:

TYPE A

This is fundamentally perceptual, where you concentrate on one aspect of the total situation to such a degree that you reject other factors in your perceptual field. Pilots become so intent on following a power line, for example, that they don't see the tower lines in the way.

TYPE B

Here, you may perceive the significant aspects of the total situation, but still be unwilling or unable to make the proper response.

Habits

These are part of our lives; many are comforting and part of a reassuring routine that keeps us mentally the right way up. Others, however, are ones we could well do without, but the trouble is that they can be very difficult to break, because the person trying to break them is the very person trapped by them. We learn habits as children, simply in order to survive. Despite our true nature, we quickly find out that if we want food, attention or "strokes", as the Americans say, we have to behave in certain ways, depending on the nature of our parents; in some families getting noticed demands entirely different behaviour than in others, mostly opposite to what we really are, which is one source of stress, the question of your true personality. In certain circumstances, habits can be dangerous, but if you can't do anything about them, we need at least to be aware of them. Training is all very well, but don't let it limit your thinking. Also, don't confuse *stereotyping* with *probability*, where you can accept a probability that certain actions will solve a similar problem to one you've had before, but stereotyping implies that the same actions work every time.

Attitudes

Flying requires considerable use of the brain, with observation and/or reaction to events, both inside and outside the aircraft. Psychology and aviation have been used to each other for some time, and part of why accidents happen is that some people are accidents waiting to happen! This depends on personality, amongst other things, and we will look at this shortly. However, personality is not the only factor to be aware of on the flight deck. Status, Role and Ability are also important. Having two Captains on board, with neither sure of who's in charge can be a real problem! Either they will be scoring points off each other, or be too gentlemanly, allowing an accident to happen while each says "after you, Nigel". How do you sort out the mess if you have someone in the left seat who is a First Officer pretending to be a Captain, and someone in the other seat who is a Captain pretending to be a First Officer?

WHAT TYPE OF PERSON IS A PILOT?

Having decided what product we are selling (safe arrival), we can now talk about the best kind of person to produce it. We certainly have more intelligence than the average car driver. Or do we? Passing exams doesn't mean you're capable of doing a decent job or handling a crisis. There are stupid solicitors, professors, you name it. I have flown with 17,000-hour pilots who I wouldn't trust with a pram, and 1,000-hour types with whom I would trust anything.

I think it's fair to say that the public typically think of pilots (when they think of them at all) as outgoing types, often in the bar and having a lark, an image that has come about from all those World War II movies. To be fair, if you were cold, hungry, tired, frightened and inexperienced, you would probably behave that way, too, but life today is quite different.

I think a pilot should be a synthesis of the following headings:

- *Meticulous* - being prepared to do the same thing, the same way, every time, and not get bored, as that's the way you miss things.

- *Forward Thinking* - in just the same way that the advanced driver is ready to deal with a corner before going into it, the advanced pilot knows that the load underneath will carry on if the helicopter slows down, and positions the controls as best he can. Unfortunately, this ability only comes with experience, but it's never too late to start.

- *Responsible* - the "responsible position" that you hold as a commander is one where you act with minimum direction but are personally responsible for the outcome of your activities. In other words, you are responsible for the machine without being directed by any other person in it.

- *Trustworthy* - people must be able to *trust* you - all of aviation runs on it. You trust the previous pilot not to have overstressed the machine, or to really have done 4.3 hours and not 6. Signatures count for a lot, and, by extension, your word.

- *Motivated.* Motivation is a drive to behave in a particular fashion. It is an internal force which can affect the quality of performance, particularly with decision making - you must be motivated to go through the process and act on the results.

WHAT IS COMMON BETWEEN COMPETENT PEOPLE?

- *Intelligence*

- *Personality.* This can be defined as "The sum total of the physical, mental, emotional and social characteristics of an individual". Generally, to be accident prone, you are either under- or overconfident. With the former, situations will tend not to be handled properly, and with the latter, situations not appreciated adequately. You might also be aggressive, independent, a risk taker, anxious, impersonal, competitive, and invulnerable, with a low stress tolerance, which, when you think about it, are all based on attention-seeking and fear. However, where personality really counts is during interactions with other people; behaviour tends to breed behaviour. Crews are frightened to deal with the Captain, and Captains won't deal with crews.

- *Leadership vs teamwork.* Leadership has been defined as facilitating the movement of a team toward the accomplishment of a task, in this case, the crew and the safe arrival of their passengers. This is a better definition than "Getting somebody to do what you want them to do" which implies a certain amount of manipulation, something more in the realm of management as a scientific process. A Leader, as opposed to a Manager, is a more positive force, inspirational, nurturing and many other words you could probably think of yourself.

- *Personal qualities* to passengers and colleagues.

UNDESIRABLE ATTITUDES

On top of *personality traits*, which you are born with, the accident-prone person has undesirable *attitudes*, which are acquired. For example, what you do (or don't do) about a bolt on the floor next to your aircraft says a lot about your personality. These traits have been identified as undesirable:

- *Impulsivity.* Doing things without forethought - not stopping to think about what you're doing, or feeling that you have to be doing something. Apply your training! The antidote is to slow down and think first.

- *Antiauthority.* These people don't like being told what to do ("Don't tell me!") They may either not respect the source of the authority, or are just plain ornery (with a deep source of bottled-up anger). Very often there's nothing wrong with this - if more people had questioned authority, we wouldn't have had half the wars, or we wouldn't get passengers pressurising pilots to do what they shouldn't. However, regulations have a purpose. They allow us to act with little information, since things are meant to be predictable, although that doesn't mean that the rules should blindly be obeyed - breaking them sometimes saves lives. The DC10 that had an engine fall off during takeoff could have kept flying if the nose had been lowered a little for speed, instead of being set at the "standard" angle of 6°, as per the simulator, which stalled the aeroplane. An official example of antiauthority is a pilot who neglects to renew medicals or ratings, or maintain records and logbooks, but my own opinion is that there's an element of laziness in there as well. The real antiauthority person is the one who keeps ignoring the Chief Pilot's instructions. The antidote is to follow the rules (mostly!)

- *Invulnerability.* People like this think that nothing untoward can happen to them, so they take more risks, or push the envelope - humility is the antidote, or the realisation that it *could* happen to you. One instructor I know cures a lot of people who insist on flying their VFR helicopters in near IMC conditions by taking them up into cloud (in a twin) and showing them how incapable they are of instrument flying, even though they can do the occasional turn with the foggles on. The point is taken! Repetitive tasks must be done as if they were new every time, no matter how tedious they may be - you can guarantee that the one time you don't check for water in fuel, it will be there!

- *Macho* people are afraid of looking small and are subject to peer pressure, which means they care too much about what other people think of them, leading to the idea that they have a low opinion of themselves, so they take unnecessary chances for different reasons than so-called Invulnerable people, above. These are typically the high-powered intimidating company executives who have houses in the middle of nowhere with no navaids within miles of the place. Such people may subconsciously put themselves in situations where they push the weather to test their own nerve or affirm their own belief of themselves. You have to learn to stick up for yourself, with management and passengers. The antidote is not to take chances, or think you can fix things on the fly.

- *Resignation.* The thought that Allah will provide is OK, but the Lord only helps those who help themselves - you've got to do your bit! If you want help to win the lottery, buy the ticket first! Resigned people do not see themselves as being able to make a great deal of difference in what happens to them - anything good or bad is down to luck. The antidote is to realise you *can* make a difference, or to have more confidence in your abilities.

As you can imagine, each side of each coin above is as bad as the other - we should be somewhere in the middle, with a possible slight bias towards antiauthority (you don't want management or customers putting you in invidious positions, and neither do you want them trying to kill you). The interesting point about the traits above is that your personal makeup of them can change from day to day.

One way of controlling hazardous personality traits is to keep a tight hold on the factors in this mnemonic:

> **S**tress
> **W**eather
> **E**xposure To Risk
> **A**ircraft
> **T**ime Constraints

Pilots must learn to avoid some classic behavioural traps:

- *Peer Pressure*, which prevents evaluating a situation objectively

- *Mind Set.* Allowing expectations to override reality

- *Get-There-Itis.* This is actually a fixation, which clouds the vision and impairs judgment, ombined with a disregard for alternative action

- *Duck-Under Syndrome.* Sneaking a peek by descending below minima, related to descending below MORA

- *Scud Running.* Trying to maintain visual contact while trying not to hit the ground - or going VFR when you really should be IFR

- *Getting Behind the Aircraft.* Allowing events to control your actions rather than the other way round, leading to.........

- *Loss of Situational Awareness*

- *Getting Low On Fuel*

- *Pushing the Envelope.* Exceeding design limitations in the belief that high performance will cover overestimated flying skills, or relying on manufacturer's fudge factors to go overweight

- *Poor Planning*

Flight Deck Management

This includes you; what sort of personality do you have? Do you leave everything to the last minute? Are you placid, or nervous and anxious? Do you have a low self-image or are you on the arrogant side? Do you succumb to pressure? Are you strong enough to stand up to the Chairman of the Board who insists he must get there NOW? Or that, whenever he is on board, you will shut down the engines immediately, rather than wait for the standard two-minute rundown period? One set of customers implied that it was the pilot's fault if they didn't get to a meeting where they stood to lose $50,000.

It's better to ask for help and look stupid, than not to ask and risk looking worse. Unfortunately, the ability to laugh at yourself and not feel uncomfortable when you've cocked things up only comes with some maturity. As you get older, you accept that mistakes are made; there's no shame in that, even the most experienced pilot couldn't fly at one time - the trick lies in not making the same mistake twice, or at least ensuring that the ones you make aren't the fatal ones.

There are so many tasks in a typical flight deck environment that they cannot all be done by one person effectively, hence the need for *delegation*, or *prioritisation* in the single-pilot case, which implies that someone must set the tasks to be performed and monitor them. That person is the Captain. With that position comes responsibilities, not least of which is the need for humility, as when recognising that someone else may actually be better than you at doing what's required, and accepting that they may do it in a different way. Also, as previously mentioned, there is the need to be a leader, and motivate, rather than drive, people - in other words, set tasks and objectives, but not necessarily the way they are done. Don't forget to provide positive support (if you're a First Officer, regardless of what your company says, it's your *job* to check on the Captain and mention it if you think there's anything wrong, which is not to say you should always be on the Captain's back - he is still the Boss. Overly chatty FOs are actually a flight safety hazard as they can stop the Captain thinking through a problem properly).

An effective leader should perform certain functions, including *regulating information flow, directing activities, motivation* and *decision making*. *Synergy* allows the group to appear greater than the sum of its parts, or crew members, in our case. It's a new name for an old concept - for example, Montgomery wrote this years ago:

> *"The real strength of an army is, and must be, far greater than the sum total of its parts; that extra strength is provided by morale, fighting spirit, mutual confidence between the leaders and the led and especially with the high command, the quality of comradeship, and many other intangible spiritual qualities."*

Groups

Many decisions are made collectively, particularly in families. In theory, therefore, a more cautious element should be built in to the process. In fact, group decisions are *more extreme* than those of the individual, meaning that an inclination to be cautious or risky will be increased in a group. This is the *group polarisation effect*.

Thus, groups are more likely than individuals to make decisions concerning risk, which is known as *risky shift*. In other words, if you put known risk takers together, the chances of them taking risks are amplified out of all proportion. *Conformity* concerns the likelihood of an individual to go along with a group decision, even if it is wrong, and it has been proved that the situations people are in, rather than their native personalities, make them behave as they do (prison guards, etc). In the single-pilot case, such an influence would come from peer pressure.

LEADERS & FOLLOWERS

Leadership has been defined as any behaviour that moves a group closer to attaining its goals. It can arise out of personality (such as a born leader), a situation, or the dynamics of the group itself (interaction).

A leader's ideas and actions influence the thoughts and behaviour of others. Through example and persuasion, and understanding the goals and desires of the group, the leader becomes a means of change and influence. Leader quality depends on the success of the relationship with the team.

There is a difference between *leadership*, which is acquired, and *authority*, which is assigned - although, optimally, they should be combined - the authority of the Captain, or Chief Pilot, should be adequately balanced by assertiveness on behalf of the crew, or crews. A follower's skills should be exercised in a supporting role that does not undermine the leader. Monty also wrote this:

> *"The acid test of an officer who aspires to high command is his ability to be able to grasp quickly the essentials of a military problem, to decide rapidly what he will do, and then to see that his subordinate commanders get on with the job. Above all, he has got to rid himself of all irrelevant detail: he must concentrate on the essentials, and on those details and only those details which are necessary to the proper carrying out of his plan - trusting his staff to effect all the necessary co-ordination."*

Out of the military context, it is equally applicable to flight crew. Just remember - leadership has nothing to do with management!

Assertiveness

The feeling of being in control, with resultant self-esteem, has health implications - a study of civil servants found that the death rate of those in lower status jobs was three times higher.

Everybody has a personal space around them, one which includes thoughts and attitudes, and culture. In other words, maintaining an appropriate emotional distance is just as important as maintaining a physical one. You do this by not putting people down, or asking too many questions, offering unwanted advice, swamping them with affection, etc. In short, allowing somebody, including yourself, to be their own person. To see what I mean about personal space, sit down next to someone on a park bench. There will be a point beyond which you do not feel comfortable going past. Only when another person sits on the bench do you feel able to bunch up closer (there is also a respectful distance behind cars).

Thus, there are many opportunities for others to invade your space, and you often have to defend it. This is known as being *assertive*, which should not be confused with being *aggressive* - assertion is a way of defending your space non-destructively. People who allow others to invade their space are *submissive*. Submission brings less responsibility and conflict, of course, and usually brings more approval, but many people use it as a form of control. Nevertheless, very often, the most appropriate role for a Captain is the submissive one, in order to get the job done.

Of course, asserting yourself with a customer or management can lead to a loss of job! Or getting seven bells kicked out of you! This is a real event:

> *"Had the opportunity to work with some real bone headed customers once - anyway, it was blowing a gale just after we had moved into a new campsite, raining also. The camp boss says its time to sling the gear off the beach up to the campsite, around 1/4 of a mile. My suggestion of waiting for the wind to die down falls on deaf ears and the comment to me was: "If you don't want to sling it up here you can go down and start carrying it up here". This after bending over backwards for these jerks for the last 30 days, so I told him if things go bad the load will be punched. His reply: "Whatever."*

> *Well I tell them the loads will have to be light as the pick up point is upwind of the drop off site, and short of hovering backwards for a 1/4 mile, turning downwind will be some fun. Well, two loads go as reasonable as one can imagine with the wind, rain and pilot with a bad attitude - now for the last load - and its just a beauty - plywood on the bottom and assorted junk on top. Unknown to me at the time was my own personal toolboxes (2).*

> *Well, getting it up was a bear, with maybe 10 torque to spare. I try to hover sideways with it...no luck...drags me to the ground....going all the way backwards was a good way to run into something near the camp, so let's try flying this crap and doing a mile wide turn. She gets going okay and into translation okay - up to 30 mph - okay - start do the wide turn okay now going downwind like a bat out of hell, trying to push enough cyclic to keep her flying. No way! The airspeed hits 0 and we start to sink - the torque is at 100%, so as calm as one can be at this moment in time, I know*

that this load is not long for the world. Try turning into wind, no way will she turn - heading for the ground - left hand ready for the emergency release in case the electric one fails, going to wait for the last minute before punching it off - about 10 feet from the ground - bombs away, gain control of old Betsy again, calmly park her near the camp, shut down, blades are still turning when the camp boss and his cohorts come over and now are saying that I did that on purpose and want to kick the crap out of me.

Well, seeing as they outnumbered me 6-1, I calmly announced that if they should like to join into fisticuffs, get in line! As I drew a line in the dirt, I said I would gladly oblige them - one at a time. After no one took up my offer and I tried explaining what I had told them before this dog and pony show got started, they shuffled off and I went over to see what had become of the bomb load. To my displeasure I found what was left of my toolboxes....flattened to about six inches high....

If you assert yourself wrongly, you will create a defensive response from the other person, and a request to reorganise two short words into a well-known phrase or saying. In other words, if you push, you will get pushed back, because the other person will not want to admit that they have affected your life, or are wrong. This is where a sympathetic approach from you will help - certainly allowing enough time for the other person to reply, and to listen - then make your point again. It may well take more than ten attempts to get your message across.

A bullying customer invades your space by attacking your self-belief and current values - how do you clear them out? Conflict is an *emotional* thing, and you have to defuse the emotion first to make any progress. This is because, if you remember, the fight-or-flight response puts people under pressure and they are not in a mood to listen until that is over.

First of all, treat the other person with respect. You won't need to say much - they will see your attitude towards them by your body language, eye contact, etc. Listen to the *meaning* of what they have to say. They won't appreciate a parrot-fashion repetition of their words, but if you paraphrase them, they may think that you are listening!

Refer to the *Communications* chapter for more on anger and how to deal with it.

NOTES

COMMUNICATION

Any relationship needs comunication to be successful. In fact, there is hardly any job in which it can be ignored. Lack of it can affect your physical health - in the 13th century, for example, Emperor Frederick experimented by cutting some babies off from all communication, instructing their nurses to stay silent. The babies all died.

Communication is important because customers, for instance, should know exactly what your helicopter can and cannot do or, more particularly, what you will and will not do (this is particularly important with heliskiing!) For some reason, a fork lift driver who says that only 50 packages can be carried is believed, but helicopter pilots are automatically assumed to be lying - most customers think that all you need to do is put in extra gas to lift the load.

Your ability to communicate will account for over 80% of your success in any walk of life. However, your current methods of communication have more than likely been based on responses learned through childhood, and can almost certainly be improved. What happens is that you build a facade, either for emotional protection or because you have to behave in certain ways in order to get what you need (food, comfort, etc.), which does not necessarily have anything to do with the person that you really are. From that stems the playing of games and manipulation, hurting and punishment until you grow into a full-blown control freak. To be sure, personality plays some part, but most behaviour patterns are learnt. Luckily, communication skills can be learnt, too.

Communication is defined as the ability to put your ideas into someone's head and be sure of success, or to exchange information without it being changed. Or both. Unfortunately, even under ideal conditions, only about 30% is retained, due to inattention, misinterpretation, expectations and emotions. Your team, if you have one, needs to know what you want done, especially in an emergency, and requires feedback as to progress and satisfaction of your expectations.

The ancient Greeks thought of a communication between two people as being in three parts:

- *Ethos* - character (and credibility) of an individual. Its essence is in the degree of trust in the words the listener believes

- *Pathos* - emotional content
- *Logos* - the logical content, and the least influential part of the whole process - it will only be listened to when the other two are clear. For example, you can be as correct as you wish, but if character and emotion are missing, your message will not get across

Verbal communication may be either *social* or *functional*. The former helps to build teamwork, and the latter is essential to flying, or operating, an aircraft. For a spoken or written message to be understood, the sender has to ensure that the receiver is using the same channel of communication, and language, and can make out the message's meaning. The point of words is to move thoughts from one person to another - they are meant to recreate the same thought pictures in the mind of the receiver, so it helps if they share the same background, for example.

The *channel of communication* is the medium used to convey the message. For the spoken word, this might be face-to-face, the radio or intercom. In fact, there can be *lack* of communication and *poor* communication. The former might be a young first officer who is very computer-literate, but doesn't tell you what he's doing. The latter, someone that tells you there is a problem, but not what it is.

So, communication is the exchange of thoughts, messages or information by various means, including speech. The elements of the process are the *sender*, the *message*, the *receiver* and *feedback*. The perceptions and background of people at either end may influence things - a person at one end of a radio transmission might receive "send three and fourpence, we are going to a dance" instead of "send reinforcements, we are going to advance".

EFFECTIVE COMMUNICATION
• •

This demands certain skills, in no particular order:

- *Seeking information* - good decisions are based on good information, so we need it to do our jobs effectively - particularly with reference to finding out what the customer actually wants
- *Problem solving*, especially in collaboration with other people
- *Listening* - active listening means not making assumptions about what the other person is saying, or what they really mean. You need to be patient, question and be supportive. Even low-time pilots

have opinions! As it happens, we can listen at up to 1200 words per minute, so our inability to listen is not physical, but mental. It takes practice to listen properly (especially, don't interrupt!)

- *Stating your position* - or *assertiveness skills*, which does *not* mean being aggressive! In other words, making sure the other person knows your viewpoint (ask yourself - would *you* listen to yourself?)

- *Resolving differences* - conflict resolution. Almost always, the best way to do this is ensure that the results are best for everyone concerned

- *Communication skill selection*, or how to perform the communication you need

- *Providing feedback*

Body Language

The fact that somebody isn't talking does not mean they are not communicating (some female silences can be quite eloquent!) It is said that 7% of communication is accomplished verbally, 38% by unconscious signals, such as tone of voice, and the remainder (55%) by non-verbal means, such as body language. In fact, before language was invented it was the only way to get your point across. It's certainly the most believed means of communication, since it will most likely reflect the true feelings of the person concerned.

Non-verbal communication can accompany verbal communication, such as a smile during a face-to-face chat. It may be acknowledgement or feedback (a nod of the head). It can also be used when the verbal type is impossible, such as a thumbs-up when it's noisy. Body language can be very subtle, but powerful. For example, the word *No* with a smile will be interpreted quite differently from one accompanied by a smack in the mouth, so it is important to include the context. Non-verbal communication may also include written information or notes, between pilots or the flight deck and cabin crew, but technology makes this even more important - it is the main way that systems speak to you, but its displays present data graphically. Unfortunately, the side-by side seating arrangements in the cockpit tend to lessen the effects of body language.

AURAL CLUES

These include the words themselves, how quickly they are spoken, and the sound or pitch of the voice, not forgetting the "ums" and "ahs" that people use when they are nervous. A good ploy is to stop listening to words and start listening to the tone. For example, you could emphasise each word in turn of the following sentence and have a different meaning every time:

I never said your dog was ugly

Visual Clues

Watch for facial expressions, posture and gestures (e.g. arms folded in defence, or looking at their watch while you are speaking).

Questions

Asking questions gives the impression that you are listening. Actually, the person who asks questions controls any conversation, because, once asked, the other person's mind flies to the answer, and you are in charge for as long as it takes to finish it.

Open-ended questions require an extended answer, such as how long is a piece of string? They are best for getting a conversation going. Closed questions require a specific answer, such as *Yes*, or *No*, and can be used to bring a conversation to a conclusion.

Listening

"Power" communicators have high levels of empathy. The key steps to proper listening include:

- *Listen* (with all the signals)
- *Pause*
- *Question* for clarification (How do you mean, exactly?)
- *Paraphrase*

Psychologists have a phrase, which is *Unconditional Positive Regard*, meaning that they always react in a positive and supportive non-judgmental way - they don't become angry or upset, but continue to smile and nod, which is obvioulsy why people go to them in the first place.

BARRIERS TO COMMUNICATION

These come from many sources, not least of which is speech - the words you say often have completely the opposite effect to what is intended, because they simply mean different things to different people. In addition, when people speak, the words become coded into some sort of indirect expression. This is because we grow up learning to be politically correct in order to get what we need from other people, thus hiding your real self behind some sort of language barrier and continual demands not to show emotion. For example, when a child asks questions at bedtime, the meaning behind the words is a request to stay a little longer. All too often, we take words at face value and confuse the real meanings for their presentation, if only because the real meat of any conversation tends to come at the end. Listeners have their problems, too, because people have filters through which words have to struggle in order to be understood.

Other factors may be a reluctance to ask questions, the influence of authority, and difficulty in listening, not forgetting making assumptions, and anger, described below. You, therefore, have to put people at their ease and make them think they can talk to you or ask questions.

In summary, it is difficult for humans to say exactly what is in their minds or hearts, and to listen without being distracted or distorting what is heard.

Responses that spoil communication include:

- *Judging*, which includes:

    - Criticising, or constant fault-finding

    - Name calling, or labelling - putting people into a box

    - Diagnosing

    - Praise, when used to manipulate

- *Solution-giving*. If you are the sort of person that always takes over, the message you are really sending is: "You are incompetent. Let me do it. I know what I'm doing." This will always tend to undermine self-esteem. Variations on this theme might include:

    - Ordering/Coercing

    - Threatening

    - Moralising. Mothers, especially, are good at this and will often introduce guilt into the equation

- Interrogation (excessive questioning)
- Advising - another self-esteem reducer
- *Avoidance*
  - Diverting. Bringing the conversation round to your own concerns, or focussing attention on yourself
  - Logic, which keeps others at an emotional distance
  - Reassurance, when it makes people feel stupid

If one party to the process is under stress, the more positive responses above will play their part in reducing communication.

Anger

When you are angry your body pumps out adrenalin, and cortisol, which depresses your immune system, so being angry can have long-term health effects. Although "losing it" in a grand firework display can make you feel better, it is only temporary and a huge exhaustive low follows as the hormones leave your system. You'll probably also have to sort out the mess with the other people! Aggressive people are more susceptible to heart attacks, clogged arteries and higher cholesterol. However, anger is also an effective means of blocking communication. This is because there are four types of angry person, each with their own language:

- those who are generally non-malicious, whose anger is quick to boil and just as quick to dissipate
- those who are slow to anger, but keep a list of everything you did or said wrong since 1929
- those who just like being angry
- those who relish the after effects rather than the argument itself

If you learn more about your own makeup, you will be in a better position to avoid setting other people off. It will help you step back and *resolve* a conflict, if you can't avoid it in the first place.

The benefits of anger include getting your own way, respect, and a defense mechanism. It is a physical reaction, whereas hostility is an *attitude*, and aggression is an *action*. Anger has four elements, arising from the body, the mind, the situation and learned responses. Anger held in becomes resentment - the trick is not to express it destructively.

If you remember, the body is not built to be in fight or flight mode as much as it is these days. You will therefore not be surprised to hear that it takes very little effort to trigger off an angry reaction after even the most trivial event. Such emotional triggers can easily make people explode, but, at the very least, will affect the way you assess situations and react to them, or, more importantly, make decisions. Emotions carry so much force and influence that they will rule your actions before you calm down enough to think rationally.

As for health, one study at the Ochsner Clinic in New Orleans reports high levels of hostility in many heart attack victims, who also had higher levels of weight, cholesterol, anxiety and depression. Stress brought on by rage can also affect memory, creativity and sleep. Bacterial infections can increase during angry episodes, and you lay yourself wide open to upper respiratory problems, like flu.

People overcome anger in many ways:

- Eating
- Displacement (taking it out on the dog)
- Talking (a lot)
- Exercise (kill the tennis ball)
- Writing
- Yelling and screaming
- Swearing
- Sulking

The best way to defuse it is to acknowledge the other person's reasons for being angry, because it is, at bottom, a frustrated demand for attention. In fact, to be effective, a display of anger must satisfy *all* of the following:

- It must be directed at the target, with no retaliation
- It must restore a sense of control or justice
- It must result in changes of behaviour or outlook (or both)
- It must use the same language (see above)

Otherwise, it will be completely non-productive. For the best results from any conflict, everyone needs to feel they are a winner (because the loser is still wound up). In our case, our customers need to come off best, and it's up to us to make them think they are, even if they're not.

BEHAVIOUR STYLES

Here are some suggested behavioural styles:

- *Assertive.* These people have respect for themselves and others and are not afraid of sticking up (politely) for themselves. Being assertive is not the same thing as being aggressive

- *Aggressive.* These people have no respect for other people, and have no problem expressing their anger, although they will blame others for it

- *Passive.* These people have no respect for themselves and excessive respect for others. They very rarely stand up for themselves

- *Passively Resistant.* These are passive people who actually try to stick up for themselves, but they use manipulative games to do it, because they still have to learn deal with people up front (watch their body language)

- *Indirectly Aggressive.* These people use underhanded methods to get their way, such as by doing jobs improperly so they won't get asked to do them again, or by using backhanded sarcastic comments, the silent treatment or gossipping, etc.

- *Passive Aggressive.* These people feel one way, but act in another - they might look happy when they are seething inside. They deny anger because it makes them feel powerless.

STUCK IN THE MIDDLE

People trying to get their way may try to put you off-balance in many ways (this happens in interviews as well). Mothers, especially, use guilt, but, in the aviation world, you are most likely to come across bullying customers or management.

Bullies choose people who will go to great lengths to avoid conflict, typically a low-time pilot in a first job who doesn't want to lose it. The problem is that, somehow, these people seem to sense your vulnerability. Unfortunately, your behaviour can make it worse, and it is the only thing you have any control over.

If you behave negatively, the other person has control of the situation, which is why they do it! Your negative reaction is their expected response, so if you do something different, by asking them why they are angry, for example, it puts *them* off balance! You can then try to direct the situation the way you want it.

A long-term remedy to this is to increase your self-esteem. Again, this can be sensed by other people, and will go a long way towards nipping awkward situations in the bud. Lt Clifton James was used as a double for Monty during WWII, and he wrote that it wasn't until he became Monty inside (i.e. with no visible change on the outside) that the whole thing began to work.

Use the following steps in any altercation:

- STOP! Remain calm and don't react with gut feelings

- Defuse the situation

- Ask questions that give the impression you really care - acknowledge their anger or concerns, fix the problem if you can

Laughter or humour is a good defuser of anger, as is reminding yourself that it won't matter in a week or so anyway.

NOTES

Life, The Universe & Everything

This may seem to be a strange chapter to have in a book like this, but people in aviation tend to have a one-sided view on things and often forget that the rest of the world exists at all (especially with aviation law). One reason why younger pilots are at such a disadvantage against bullying customers or management is that they lack life skills, or a conception of themselves in the great scheme of things (the helicopter industry, in particular, seems to have a mass self-esteem problem).

Another thing that I have aways been curious about is where that "gut instinct" comes from - the intuition that tells you that the hole in the trees is too small for your helicopter, or that the runway is too short for takeoff. We hope to explore the still, small, voice in this chapter and maybe help you use it better (the Captain of the Gimli Glider, the 767 that ran out of fuel near Winnipeg, had an uneasy feeling in Ottawa that there was something wrong with the fuel load, but everything checked out and he ignored it. The mechanic's van ran out of gas, too, on the way to rescue it!)

It was noticed, after the Asian tsunami, that no animals died, and primitive tribesmen joined them in their flight to higher ground *well before it happened*. The US *Journal of Meteorology* reports that two horses were quietly grazing in a field when they suddenly bolted to the far end. Around fifteen seconds later, a lightning bolt struck where they had been originally standing.

An area of the brain near the frontal lobes (along the walls that divide the left and right hemispheres) that monitors environmental clues has been identified by Washington University in St Louis, that works at a subconscious level. Drs Joshua Brown and Todd Braver point out that activity in this area has been documented when people make difficult decisions after making a mistake, and it is also possible that a mistake can be recognised before it is made.They proved this by monitoring brain activity with fMRI while people pushed buttons on coloured cues. It was found that they figured out by themselves that one colour was more likely to be wrong (naturally, they were not told of the colours' relationship).

IN TWO MINDS

The first thing to realise is that there are other things in this world than the normal things we see and hear. Dogs can hear sounds that are outside the range of our ears, and our eyes don't see everything, either. The human body is actually made up of two parts, one you can see, and one you can't, the latter usually being completely forgotten about, even though it has the greatest influence on our lives or actions. The two parts are not supposed to work separately, yet they have completely different characteristics and are often at odds with each other. I am talking about the conscious and subconscious (unconscious) minds, and if you ever needed proof that our bodies are host to more than one life, consider that, after we die, our nails and hair keep on growing for several days.

Any psychologist will tell you that we are slaves to our subconscious, and the advertising industry spends millions each year in proving them right (two books are worth reading in this respect - *The Hidden Persuaders*, by Vance Packard, and *Subliminal Seduction*, written by somebody whose name escapes me right now. However, they were both written a long, long time ago, and are still more than valid!) Anyhow, advertisers use techniques that access your subconscious mind directly, by going past the conscious and its reasoning power, so you can't block it. As a result, we are triggered to respond to certain events, such as: Father's Day - open wallet - spend money (such occasions have been invented by the retail industry. Mother's Day, based on Mothering Sunday, is about the only one that has any basis in reality).

The unconscious mind contains a record of all that we have ever done, seen or heard, but it is not immediately accessible by the conscious mind - there is a barrier between the two. You could liken it to a filing cabinet, which is very good at remembering things, but has no reasoning power - that is the job of the conscious, which usually cannot remember much at all (I refer you to the difference between short and long-term memory in Chapter 2). It is your subconscious that determines your body language, tone of voice, etc. when interacting with other people, and it is their subconscious that picks up the same clues for their own information. *You even tell your own subconscious who you are!* If you don't, the world, in the shape of other people, such as customers, will do it for you.

Around your body, there is what can best be termed a magnetic field, which is hardly surprising, as the body depends on its own self-generated electricity to function (this is why we need metal elements in our diet to react with acidic bodily fluids to keep it going). Thus, there is, by definition, a magnetic field, since all electric currents have one.

As previously mentioned, you can prove the existence of such a field by taking the average park bench and sitting down on it. When somebody else comes to sit down, they will keep a respectful distance away. It is only when another person comes along that they feel they have "permission" to get closer to you. Similarly, all shoolchildren know the trick of running their hands over the top of somebody's head and pretending that an egg has just been broken - their hands don't touch anything, but it still feels authentic. And just watch how people react when you drive too close behind them! This field around the body is where the unconscious mind lives, and it connects with those of other people, which is something that has been established by the Russians, and the US Army, in long-standing, documented, studies of the subject (*Reading The Enemy's Mind*, by Major Paul Smith). Thus, contrary to popular belief, the body lives in the mind, rather than the other way round.

In fact, everybody's unconscious mind is connected to what Jung called the "collective unconscious", which can be likened to a vast cloud of magnetic energy hovering above them, to which they are all linked:

This explains how some organisms seem to evolve all by themselves, since it has been shown that the experiences of one animal can get transmitted to others, with subsequent evolvement of the species.

The conscious mind is quite slow in comparison, as well as being less capable - some psychologists have determined that we process something in the order of several billions of pieces of information every second, only 2000 of which we are aware of. The figures sound fantastic, but the relative proportions would seem to be about right, in that 95% of the brain's activity does not relate to conscious reality. The point is, as humans, a lot of what keeps us going goes on quietly in the background, without us knowing anything about it.

UNDUE INFLUENCE

Everything starts with the way we think. Thinking is an electrical activity, which can be recorded with an ECG machine. Thoughts of a like nature will congregate in their own part of the unconscious mind (remember that it doesn't forget anything) and occupy a increasing part of it, until they eventually gain an influence all of their own. Thus, if you continually worry about your weight, for example, a part of your subconscious will become "top heavy" with all those thoughts and begin to attract events and people into your life (through the collective unconscious) that will reinforce this image - the unconscious mind has no reasoning power, so is unaware that this could be potentially harmful. Thus, if you constantly worry about your weight, you will gain a concentration of thoughts about it in your subconscious mind, and you can diet all you want, but the influence gained by this collection of thoughts will undo all your good work. Similarly, if a low self-image is reinforced by other people, and you accept their view, you will have a heightened sense of your own inferiority

This is what psychologists call a complex. Also, if you focus on not smoking, all you are really doing is bringing smoking into the attention span - a better tactic is not to think of it at all! The obvious way of getting out of such patterns is to create remedial thinking of an opposite nature, but that is easier said than done, and is the subject of another type of book.

That Still, Small Voice

The unconscious mind, which can process data far faster than the conscious mind, will try to get its message through in many ways, the most obvious example being dreams, but in order to get it to help us in our daily work, we have to think of an alternative method.

It is possible to develop the ability to listen to what is inside - the way I did it was to have somebody place an object on their desk, ring me up and get me to describe what it was. The objects ranged from a pencil to a rolled up ball of paper. At first, all you do is guess, or try to think too hard, but after a while, you develop the ability to *feel* what the answer is, in the solar plexus area (which is what you are supposed to listen to opera with).

Once you get to recognise the feeling, you know with some certainty that your answer will be right, without necessarily knowing the reason why. You will be very surprised at your success rate - the essential trick is learning to believe the first thing that comes into your head, even if it sounds ridiculous. This ability helps me most when operating low level with little time to look at a map - all of a sudden I get an urge to look at it, and, sure enough, there's something I need to avoid. It works with parking spaces, too. The other thing I have noticed that I always seem to ring people just before they ring me - I should learn not to, so I can make the phone bill cheaper!

However, your subconscious can be "primed" to make you think or behave in certain ways, as mentioned before. Some elderly people did far worse on memory tests when words associated with senility were flashed before them so quickly that only their subconscious minds could recognise them. Similarly, other people were subdued into waiting patiently after completing a word puzzle with cues about politeness, courtesy and consideration. If you are frustrated and under stress before taking any tests or being in any situations, you will perform less well than if you were cheerful. Well, you would, wouldn't you?

This only serves to reinforce the points made in previous chapters about proper responses to situations being affected by your beliefs and predispositions. In fact, the brain will edit information it receives to fit preconceived beliefs (around 50%), so you should always try *not* to see what you want to see. Anything too fantastic, such as a bad childhood, will be switched out. In Quantum Physics, for example, they have determined that particles take 2 forms - they can be waves or particles. A wave will become a particle only when it is observed, or perceived by the brain.

SUMMARY

You can make decisions based on intuitive feelings (what some people might call "snap judgements", but you must learn to recognise the likely causes of error.

What is around you (your world) reflects directly on who you are inside. Thus, what you believe determines what you make true, and anything that goes against this can be modified by the brain. "Our light is what our thoughts make of it" - *Marcus Aurelius.*

INDEX